World Scientific Lecture Notes in Complex Systems – Vol. 3

editors

Dieter Armbruster
Arizona State University, USA

Kunihiko Kaneko
University of Tokyo, Japan

Alexander S. Mikhailov
Fritz Haber Institute, Germany

Networks of Interacting Machines

Production Organization in Complex Industrial Systems and Biological Cells

World Scientific

NEW JERSEY • LONDON • SINGAPORE • BEIJING • SHANGHAI • HONG KONG • TAIPEI • CHENNAI

Published by

World Scientific Publishing Co. Pte. Ltd.

5 Toh Tuck Link, Singapore 596224

USA office: 27 Warren Street, Suite 401-402, Hackensack, NJ 07601

UK office: 57 Shelton Street, Covent Garden, London WC2H 9HE

British Library Cataloguing-in-Publication Data
A catalogue record for this book is available from the British Library.

NETWORKS OF INTERACTING MACHINES
Production Organization in Complex Industrial Systems and Biological Cells

ISBN 981-256-498-5

Printed in Singapore by World Scientific Printers (S) Pte Ltd

PREFACE

This book is devoted to a discussion of analogies and differences of complex production systems – natural, as in biological cells, or man-made, as in economic systems or industrial production. Hierarchical models of industrial supply chains are characterized by systems of differential equations closely related to traffic flow models. Interesting questions asked in this context go much beyond the static network analysis. Since most production systems are highly stochastic and event-driven, they are rarely, if ever, at equilibrium. Hence the reaction of a given supply network to specific events as well as its transient response to parameter changes and control actions has to be understood through simulation as well as through fundamental models. Additionally, these dynamic responses are constrained by economics, leading to the evolution of systems with higher productivity and profitability. On the other hand, recent research has put forward a view of a biological cell as a factory where products of one machine are used by other machines for manufacturing of their own products or for regulation of their functions. In some cases, a cell may operate in a synchronous mode, so that the operation cycles of individual machines are temporally correlated and the intermediate parts are released exactly when they are needed for further production by other machines. A special property of biological production networks is that, to a large extent, their activity is self-organized and persists despite the presence of strong thermal fluctuations. Progress in micro- and nanotechnology may soon bring about a possibility to produce nanorobots and to design self-reproducing artificial cells, leading to a revolution in industrial manufacturing. However, the principles of purposeful operation of large ensembles of nanodevices and populations of artificial cells should first be investigated. Obviously, they again should bear strong similarities to biological organization. The common themes of industrial and biological production include evolution and optimization, synchronization and self-organization, robust operation despite high stochasticity, and hierarchical dynamics.

The book presents selected lectures given at the international workshop "Networks of Interacting Machines: Industrial Production Systems and Biological Cells" (Berlin, December 2003), organized with the financial support of the Klaus Tschira Foundation. Its authors are a group of scientists and industrialists from Europe, Japan and USA. Together, we hope to provide an overview of modern perspectives on principles of production organization. We are grateful to the Tschira Foundation for providing financial support and thank Dr. Oliver Rudzick for his assistance in the preparation of manuscripts.

Dieter Armbruster
Arizona State University
Tempe, AZ, USA

Kunihiko Kaneko
University of Tokyo
Tokyo, Japan

Alexander S. Mikhailov
Fritz-Haber-Institut der Max-Planck-Gesellschaft
Berlin, Germany

CONTENTS

CHAPTER 1

CONTINUUM MODELS FOR INTERACTING MACHINES

Dieter Armbruster[1], Pierre Degond[2], Christian Ringhofer[1]

(1) Department of Mathematics, Arizona State University
Tempe, AZ 85287-1804, USA
E-mail: armbruster@asu.edu, ringhofer@asu.edu
(2) MIP, UMR 5640 (CNRS-UPS-INSA), Université Paul Sabatier
118, route de Narbonne, 31062 Toulouse cedex, France
E-mail: degond@mip.ups-tlse.fr

A review of continuum models for production flows involving a large number of items and a large number of production stages is presented. The basic heuristic model is based on mass conservation and state equations for the relationship between the cycle time and the amount of work in progress in a factory. Heuristic extensions lead to advection diffusion equations and to capacity limited fluxes. Comparisons between discrete event simulations and numerical solutions of the heuristic PDEs are made. First principle models based on the Boltzman equation for a probability density of a production lot, evolving in time and production stages are developed. It is shown how the basic heuristic model constitute the zero order approximation of a moment expansion of the probability density. Similarly, the advection diffusion equation can be derived as the first order Chapman-Enskog expansion assuming a stochastically varying throughput time. It is shown how dispatch policies can be modeled by including an attribute in the probability density whose time evolution is governed by the interaction between the dispatch policy and the capacity constraints of the system. The resulting zero order moment expansion reproduces the heuristic capacity constraint model whereas a first order moment will lead to multiphase solutions representing multilane fluxes and overtaking of production lots. A discussion on the similarities and differences of industrial production networks and biological networks is also presented.

1.1. Introduction

As large factories or other large production systems have become increasingly more complicated, their dynamic behavior over time and in response to changes in the production environment is of outmost importance for attaining the overall goals of timely and cost efficient production. As a result, there is a substantial research endeavor worldwide to simulate, optimize and control such production systems. This paper will give an overview of our recent contributions to that endeavor - through continuous models and their simulation of networks of interacting machines.

Before we will go into details we will try to discuss our view of the theme of this book, in particular we will outline some of the dichotomies between naturally occurring networks of machines, especially biomolecular machines [20], and factory production.

- *Self-organizing vs. planning*: Molecular biological networks spontaneously cluster and synchronize in order to produce a desired product, say a protein. Factories production as well is highly organized and synchronized (e.g just in time production). However, rarely does this synchronization occur spontaneously through interaction with other production processes. Typically it is facilitated through production layouts, production rules and recipes and is controlled through an external management system.
- *Stochastic behavior vs. regular production:* Biological processes typically live in a highly fluctuating, unstable environment where transport may be diffusive and production efficiency is very low. Nevertheless, on average most biomolecular processes are remarkably stable. In contrast, for instance, the production of semiconductor chips which is one of today's most advanced manufacturing processes is run in the most physically controlled way imaginable.
- *Evolutionary optimality vs. profit:* Evolution is a long term randomly executed continuous search process for a optimally adjusted organism. The goal for a production system is to maximize profit over usually a short timescale.
- *Reaction kinetics, diffusion and continuous transport equations vs. discrete event models:* Often the intermediate steps and the intermediate products of many biomolecular production systems are not known and neither are the transport paths. Hence often models in the form of reaction diffusion equations and continuous transport equations are developed that either aggregate the biomolecular

production in space and/or split it into short and long timescales where only the time evolution of the long timescale is modeled. In contrast, in a typical model for production systems individual products and individual production steps can be characterized in great detail, including their stochastic behavior. The resulting models therefore tend to be discrete event simulations.

This list is certainly not exhaustive but it seems to suggest that there is not much that the biomolecular production networks and the industrial production networks have in common and that research concepts and techniques may not be successfully transfered from one realm to the other. However, a more detailed inspection of these dichotomies reveals that both areas have much more in common than meets the eye initially.

- The stability of self-organizing systems under perturbations resulting in redundancy and self-correcting behavior has been discovered in several production systems. A typical example is the case of a bucket brigade production system: The term "bucket brigade" was coined by Bartholdi and Eisenstein [9], for production lines in which the workers hand over their workpiece to the next worker down the line, whenever the last worker has finished his job. If workers are sequenced from slowest to fastest, then there is a stable fixed point that the system will converge to. Not only is this self-organized production optimal, it also selfadjust to the loss of a worker by distributing the work over the remaining workers in an optimal way. Another example has been discussed in [22] where a network of machines leads to synchronization at least for the average production rates in different production levels. Agent based models [10] have also been used to study emergent dynamic behavior in supply chain modeling. The overall concept of self-organizing production has lead to at least one large research unit that is completely dedicated to this theme [23].
- Stochasticity is far from under control in industrial production systems. Although every effort is made to have a stable and completely controlled production environment for a semiconductor chip factory, the actual production varies dramatically: For instance, figure 1.1 shows the actual time dependent path of 200 lots in an actual INTEL factory (the data have been rescaled to protect proprietary information). The resulting throughput times vary by about a hundred percent. Given that the raw throughput time through such a

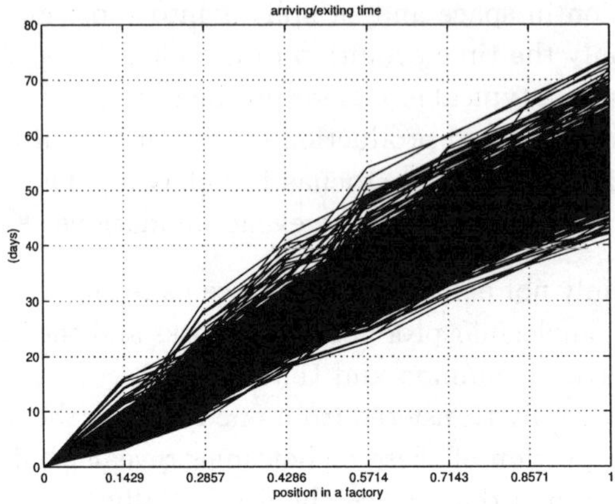

Fig. 1.1. Paths of 200 lots through an actual factory.

factory is of the order of weeks, it is clear that prediction for the time when a particular chip is leaving the factory is next to impossible, or at least has to reflect a very high level of uncertainty. Similarly, for a typical new production process for the latest and fastest chip, the resulting yields are quite small and vary dramatically between different production lots. As a result, in order to produce the high value products, many low value byproducts are produced that may or may not be commercially valuable. In that sense the uncertainty in both, biomolecular production and industrial production may be very high and similar questions pertaining to efficiency and stability are unanswered. This suggests that methods to understand, control or reduce the stochasticity in one area may well be applicable in the other.

- Biological evolution produces the blueprint and the production recipes for a successful species in much the same way as capital investment into machines and research produces the means for a successful industrial enterprise. However, in both cases, it is often very unclear how the individual production step relates to the overall evaluation function and whether a particular change will lead to an improvement. Industrial production typically generates metagoals like a high utilization of a factory, low inventories, just in time delivery etc. every one of which may contribute to the success of the whole enterprise but collectively those goals are often contradictory and often have unspecified trade-offs. It is yet completely unexplored whether evolutionary optimization and the fitness of an industrial enterprise can learn from each other.

- There seems to be a clear case for discrete event simulations or at least for hybrid simulations [12] in biomolecular systems: Many biological processes are triggered in some way, suggesting that a simulation that has many concurrent processes that are synchronized by specific events may be a useful model. Hybrid processes would be applicable in cases with different timescales - a low one evolving on a continuous description of space and time interrupted by specific events that trigger a different slow evolution. On the other hand, we have spent the last several years to develop models for production systems that are continuum based. The rest of this paper will discuss them and will show a detailed description of heuristic as well as first principle models.

1.2. Heuristic Models

1.2.1. *Quasistatic Models*

Detailed modeling of complicated production networks is usually done via discrete event simulations. These are stochastic simulations and hence many experiments are performed to generate a large dataset. Subsequently this ensemble of experiments is postprocessed to extract the desired characteristic quantities like average cycle time, average throughput and variances for these characteristic quantities. For time dependent phenomena there are two major concerns associated with that approach: i) postprocessing is not always straightforward and the detailed algorithms used may influence the results, ii) time dependent phenomena have their own timescale that typically leads to a need for larger ensembles. Hence small prototype systems can most often not be scaled up to relevant systems because the computational costs become prohibitively expensive.

One way to deal with this problem is to consider the time evolution of densities rather than individual lots. This leads to the so called fluid models which consist of networks of coupled ordinary differential equations coupled through flux conservation: The rate of change of a queue in front of a machine is given through the influx to that machine (λ_i) minus its outflux (μ_i),i.e.

$$\frac{dq_i}{dt} = \lambda_i - \mu_i \, , \tag{1}$$

where $\mu_i = \lambda_{i+1}$. Fluid models have been studied for quite some time (e.g. [13]) and it can be shown that a queuing system is stable (has finite queue lengths), if and only if the associated fluid model has a stable steady state. The major drawback of a fluid model is that it does not have good representation of the delay in a system. Any part of the production system that is modelled by an equation like Eq. (1) has an instantaneous change in outflux, if there is a change in influx. In a sense, a fluid model stops half way to a real continuum model like a fluid flow. Hence we proposed in [2] a second independent variable x which describes the degree of completion for a production process. Note that the degree of completion as the "spatial" variable x allows us to untangle any topologically complicated physical path through the machines in a factory into a simple one dimensional linear chain. Our state variable is now $\rho(x,t)$, the density of product at a stage x at a given time t. Clearly, in its simplest form the time evolution of this density has to respect mass conservation. Hence we have a partial differential equation of the form

$$\frac{\partial \rho}{\partial t} + \frac{\partial (v\rho)}{\partial x} = 0 \tag{2}$$

$$v\rho|_{x=0} = \lambda(t) \tag{3}$$

$$\rho(x,0) = f(x) \tag{4}$$

This is a conservation law with $v\rho$ the flux, the boundary condition $\lambda(t)$ being the external influx and $f(x)$ an arbitrary initial condition. Such conservation laws typically lead to a hyperbolic wave equation where the details of the dynamics have to be represented by a state equation for the velocity v, i.e. the functional form of the dependence of the velocity on the density. Typical models are

$$v_{LW}(\rho) = v_0(1 - \frac{\rho}{\rho_c}), \tag{5}$$

$$v_Q(\rho) = \frac{v_0}{1 + \frac{L(\rho)}{L_c}}, \tag{6}$$

$$v_{eq}(\rho) = \Phi(L), \tag{7}$$

with L the total load (Work in progress, WIP) given as

$$L(\rho) = \int_0^1 \rho(x,t)dx. \tag{8}$$

Equation (5) for instance, treats production flow like a traffic flow as in the traffic model of Lighthill and Whitham [18]. The interaction of the various products in the factory here is strictly local: An increase in the local density reduces the velocity at that position until, at a critical density ρ_c the velocity goes to zero. Such a model typically has shock waves corresponding to traffic jams. Equation (6) corresponds to a queuing theory model where, in steady state, the time $\tau = 1/v_Q$ to exit a queue becomes $\tau = \tau_0(1 + L)$ with τ_0 the time that a product spends in the machine without any waiting and L represents the total length of the queue, or the total work in progress (WIP) (Eq. 8). In general, the quasi-static approach using a state equation for the velocity is closely related to the concept of a clearing function [8,19] which is a static representation of the dependency of the throughput on the current WIP in the factory. Equation 7 is a generic model for such a clearing function.

The dependence of the velocity on the total WIP rather than on the local density in all but the Lighthill and Whitham model reflects our emphasis on semiconductor manufacturing. There, chips are produced in layers which are build up by cycling repeatedly through the same factory. Hence, since lots that are at different layers compete for machine time, fluctuations at the beginning of the production process can influence not only the production upstream but also the production downstream.

All state equation models are based on abstractions on the actual stochastic processes inside the network of machines. These abstractions can be parameterized using either real data or through a very detailed discrete event simulation. The latter not only represents the 'hardware'(i.e. the machines) of the production process but also the 'software' which are the production rules and policies. In any case, we can extract the functional relationship $\Phi(L)$ i.e. the clearing function (Eq. 7) from the available data.

Figure 1.2 shows such a state equation resulting from a detailed computational experiment. The simulated network consists of 5 machines and is re-entrant, i.e. the production recipe requires that each lot will have to go through the 5 machines four times before it exits. We see that a linear fit as in a queuing model will be quite appropriate. Clearly such a state equation can be interpolated between a few detailed simulations and then used to predict the outflux or the throughput time in steady state for an influx that has not been simulated. That is the way that static clearing functions models are used. It does not need any dynamics as represented in the PDE. The PDE however, allows us to study transient behavior like for instance a step up or down of the influx into a factory. The continuity equation (Eq. 2) with a steady state model for the velocity represents a quasi-static or adiabatic model: any fluctuation in the WIP-level leads to an instantaneous relaxation of the velocity to the steady state velocity given by the state

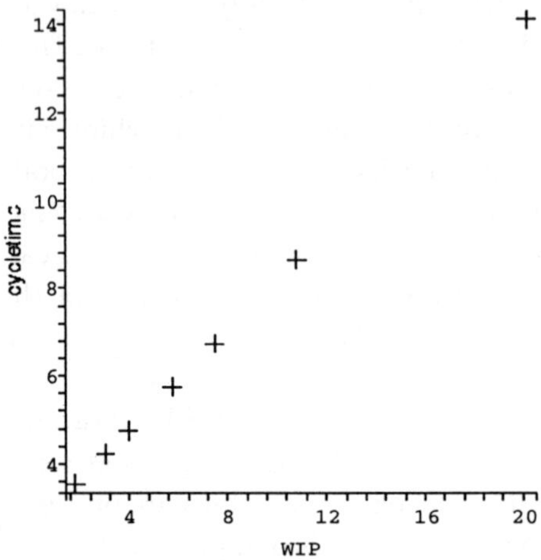

Fig. 1.2. Seven datapoints for a state equation describing the relationship between cycle time and WIP.

equation. Figures 1.3 and 1.4 show such an experiment: For the factory model that gave us the steady state relationship in Figure 1.2 we performed a step-up experiment from an influx of $\lambda = 1/0.8$ to an influx of $\lambda = 1/0.7$. The noisy curves in Figs 1.3 and 1.4 are the averages for the throughput and the cycle time of a large number of discrete event simulations of that experiment. The smooth curves represent the quasi-static PDE simulation using the state equation derived from Figure 1.2. While the transient is not perfectly resolved, the agreement is not bad either. The large downward spike in the throughput for the PDE simulation results from the fact that in our model the velocity is spatially uniform and depends on the total WIP in the factory. Hence any increase in WIP (through e.g. an increase in influx) will lead to an instantaneous reduction in velocity and hence to an instantaneous reduction in outflux. Obviously, while the discrete event simulation has some of this features, it is not re-entrant enough for such a strong reaction.

1.2.2. *Advection-Diffusion Equations*

In analogy to stochastic particle or fluid transports one would expect that the next higher order effect beyond pure advection would involve diffusion. The resulting differential equation is an advection-diffusion equation of the

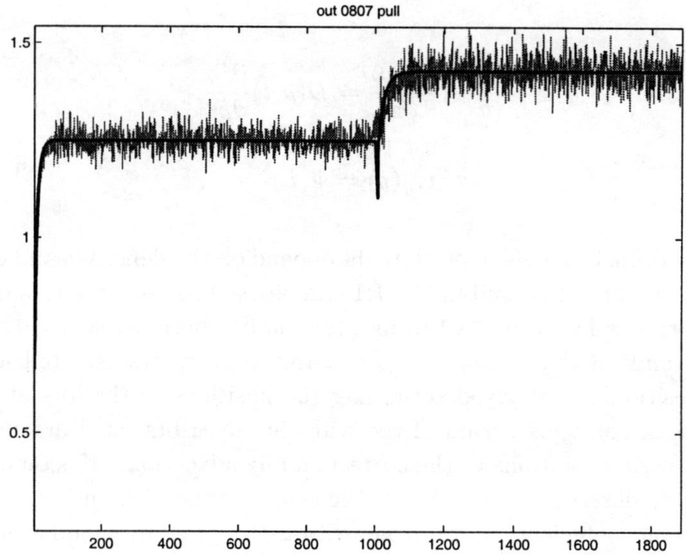

Fig. 1.3. Throughput as a function of time for a step up in input at $t = 1000$.

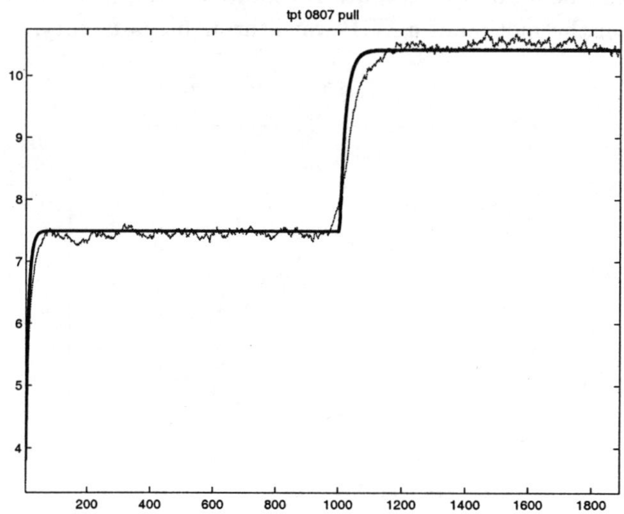

Fig. 1.4. Cycle time as a function of time for a step up in input at $t = 1000$.

form

$$\frac{\partial \rho}{\partial t} + \frac{\partial (v\rho)}{\partial x} = D(\rho, t)\frac{\partial^2 \rho}{\partial \rho^2}, \tag{9}$$

$$v_{eq}(\rho) = \Phi(L), \tag{10}$$

where the diffusion coefficient D might depend on the density ρ and on time. The factory data displayed in Fig 1.1 can be used to estimate this diffusion coefficient: The fan of paths through the factory becomes wider over time. We can think of this figure as a δ-distribution that was started at $t = 0$ at the position $x = 0$. By determining the positions of the lots at a given time we can get a histogram of the widening distribution. Data matching of the Gaussian solutions to the corresponding advection diffusion equation allows us to determine the speed of the center of the distribution as well as the diffusion coefficient. In principle, with better resolution and more lots it would be possible to identify regions of high diffusion in the factory. With only 200 lots evaluated at less than 10 positions along the factory we can only get an order of magnitude for the diffusion coefficient. Details of this discussion can be found in [1]. Figure 1.5 shows the influence of the diffusion coefficient on the motion of a WIP-wave through the factory.

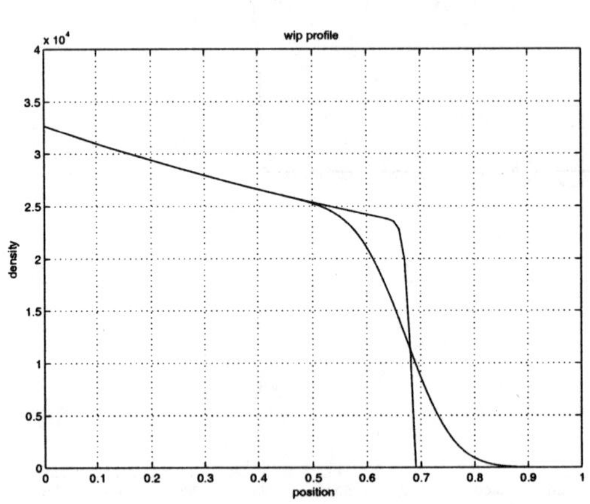

Fig. 1.5. WIP profile after a step up in influx, with and without diffusion.

1.2.3. *Policies and Bottlenecks*

So far we have treated the factory as homogeneous in the stage variable x. I.e. the velocity of a moving lot was the same at any position in the factory. While we might not want to go into the details of modeling individual stochastic behavior of machines or the production step at a certain stage (since in that case we would be back at discrete event simulations that are very time consuming), we might be interested in the influence of dispatch policies and in the behavior at or near certain important machines that represent a bottleneck. To study these issues, we can augment our heuristic models.

1.2.3.1. *Dispatch Rules*

For any topologically complicated flow where lots at different stages of the production process approach the same machine, dispatch rules are needed to decide which step the machine should do next, once it becomes free. The three major dispatch rules that are used are FIFO, PUSH and PULL policies. In FIFO the sequence of lots through the machine corresponds to the sequence of arrivals. In a PUSH policy the arriving lots are ordered according to their stage number and lots with lower stage number receive preference. In a PULL policy, lots with higher stage number receive preference. Typically a PULL policy is used for "make to order" processes where production tries to satisfy a given order at a given date. A PUSH policy typically reflects a "made to plan" process were production typically fills a warehouse. While to first order FIFO does not distinguish between stages in the factory and hence a homogeneous model will be a good approximation, PUSH, PULL and any other more complicated rule can be incorporated via integration kernels in our state equation. Let $w(x,s)$ indicate the importance of a queue at location s in completion space on the speed of a lot at location x. Then the velocity at position x can be written as

$$v(x,t) = v_0(1 - \frac{1}{L_{max}} \int_0^1 w(x,s)u(s,t)ds). \tag{11}$$

As a result, the velocity will cease to be uniform throughout the factory. For instance, a pull policy is modeled by the kernel

$$w(x,s) = \begin{cases} 0 \text{ if } s < x, \\ 1 \text{ if } x \leq s, \end{cases} \tag{12}$$

leading to

$$v(x,t) = v_0(1 - \frac{1}{L_{max}} \int_x^1 u(s,t)ds). \tag{13}$$

Hence, $v(1,u) = v_0$, indicating that product at the end of production moves independently of the load of the factory, while $v(0,u)$ shows the full impact of the loading of the factory on the motion at the beginning of the production line.

1.2.3.2. *Bottlenecks and Maximal Capacities*

There are several fundamental stochastic processes that occur in a production system: Anything involving operators will lead to uncertainty as well as unscheduled machine breakdowns and repairs. A state equation model describes the influence of these processes in a very comprehensive average. We will show in the first principle models in the next section how to do more sophisticated averages. However, there is one more heuristic extension that can be done: While it is extremely hard to characterize in any meaningful way the actual stochastic processes involved we know that there exist some physical limits. In particular, a fixed production line has a maximal production capacity at every machine. No matter what policies, the recipe for a certain chip expects it to stay x hours in a diffusion oven. Hence we can define a maximal capacity function $C(\xi)$ that describes the capacities of all machines that are involved in the production process, where ξ now is a variable that does not describe the stages but the sequence of machines. The resulting quasi-static model then becomes

$$\begin{aligned} \tfrac{\partial \rho}{\partial t} + \tfrac{\partial F}{\partial x} &= 0 \\ F(x) &= \min\{\rho v_{eq}, \mu(x)\} \\ v_{eq} &= \Phi(L) \end{aligned} \tag{14}$$

where the maximally available capacity at stage $\mu(x)$ depends on the dispatch policy: Assume that there are n layers that are produced on the same machines leading to n loops through those machines. The map between machine position and production stage is then given by modular division: The stage variable x acquires a layer index i such that $x_i = \frac{i-1}{n} + \xi$ for $i = 0..n-1$ describes a production stage in the $i+1$th loop at the machine position ξ. At any particular machine, flux requests from all n loops may compete for the maximally available capacity $C(\xi)$ leading to a distribution of the maximally available capacities at stage i, $\mu(x_i)$, depending on the fluxes $F(x_j)$ in the other loops. For instance for a PUSH policy, capacities are allocated from front to end. Hence we can iterate, for $i = 0..n-1$ the following scheme to find the capacity distribution $\mu(x)$.

$$\mu(x_0) = C(\xi)$$
$$F(x_i) = min\{\mu(x_i), \rho v_{eq}|_{x_i}\}$$
$$\mu(x_i) = max\{0, \mu(x_{i-1}) - F(x_{i-1})\}$$

Such a model will lead to the formation of bottlenecks and hence δ-distributions in the density variable for any influx that temporarily exceeds the total capacity.

1.3. First Principle Models

1.3.1. *Kinetic Models*

The models described in the previous section are more or less heuristic in nature. That is, the only physical principle used was product conservation, leading to the obvious conservation law $\partial_t \rho + \partial_x F = 0$ for the part density ρ and the flux density F. The dependence of F on ρ was assumed to be given, based on specific models. In this section we will review a more subtle approach, namely to derive equations for densities of parts based on simple rules for the movement of parts themselves. Generally, whenever particles which carry some attribute y move through space x with a certain velocity $u(x, y, t)$ dependent on position and attribute, the underlying equation for the density $f(x, y, t)$ of a particle being at position x with attribute y at time t is the Boltzmann equation

$$\partial_t f + \partial_x[u(x, y, t)f] + \partial_y(E(x, y, t)f) = Q[f] \ , \tag{15}$$

where the term $\partial_y(Ef)$ models acceleration by a force E, i.e. a continuous change in the attribute, and the term $Q[f]$ models a random and discontinuous change in the attribute. $Q[f]$ will be an integral operator in attribute direction whose kernel is related to the probability distribution of the newly chosen attribute once the discontinuous change has occurred. Because of conservation of product (no parts are lost in this abrupt change), the total number of parts at position x has to remain the same before and after the change, resulting in the condition

$$\int Q[f](x, y, t)\ dy = 0, \ \forall f \ . \tag{16}$$

If we define the part density ρ and the flux density F at position x and time t by

$$\rho(x,t) = \int f(x,y,t)\, dy, \quad F(x,t) = \int u(x,y,t) f(x,y,t)\, dy\ ,$$

integrating (15) with respect to the attribute variable y gives, using (16), the conservation law

$$\partial_t \rho + \partial_x F = 0\ .$$

The advantage of the kinetic approach lies in the fact that the rules and underlying assumptions going in the formulation of equation (15) can be much more detailed than those involved in modeling F simply as a function of ρ and its derivatives and integrals. From an information theoretic point of view, the models in the previous section basically state that we can compute the flux density F by just knowing the value of ρ (and maybe some of its derivatives and integrals). In contrast, the relation between ρ and F is much more complex here, involving the density f, and implies that we need to know the whole history of ρ to compute the flux density F. We heavily use the methodology of classical gas dynamics, where equation (15) corresponds to the Boltzmann equation for a rarefied gas and the models of the previous section play the role of the basic equations of gas dynamics, the compressible Euler equations and the Navier - Stokes equations. We will refer to (15) as the kinetic equation underlying the macroscopic equations of the previous section. We refer the reader to [11] for an overview of the underlying theory. It should be noted, that, from a theoretical as well as a numerical point of view, the kinetic theory of supply chains is simpler than the corresponding theory of gas dynamics, since the stage variable x ('space') will in general be one dimensional.

1.3.2. *Deterministic Kinetic Models*

We will first address the case of a purely deterministic movement of parts through the system. This will lead to gas dynamic equations of the type discussed in Section 2. This means the resulting kinetic equation (15) will be purely hyperbolic and all information about the system at any time is given deterministically by the influx. We start by defining the motion of parts through the system via trajectories. The Newton equations of motion of part number n with position $x = \xi_n(t)$ and attribute $y = \eta_n(t)$ are given by

$$\partial_t \xi_n = u(\xi_n, \eta_n, t), \quad \partial_t \eta_n = E(\xi_n, \eta_n, t), \tag{17}$$

where the functions u and E are the velocity and the accelerating force of the particle. The choice of u and E depends on the model under consideration. We assume that the part number n enters the production system at time $t = a_n$ with an attribute $y = r_n$ and therefore impose the initial conditions

$$\xi_n(a_n) = 0, \quad \eta_n(a_n) = r_n,$$

for (17). Considering a system of N parts, the density function $f(x, y, t)$ in this picture will be given by

$$f(x, y, t) = \sum_{n=1}^{N} \delta(x - \xi_n(t))\delta(y - \eta_n(t))H(t - a_n) ,$$

where $H(t)$ denotes the usual Heaviside function, denoting that the part does not exist in the system for $t < a_n$ Since f is a measure, equation (15) can of course only be valid in its weak form, which is obtained by multiplying (15) with a compactly supported test function $\psi(x, y, t)$ and integrating by parts.

$$\int_0^\infty dx \int_0^\infty dt \int dy\ f(x, y, t)[\partial_t\psi + u\partial_x\psi + E\partial_y\psi] = \tag{18}$$

$$\int_0^\infty dt \int dy\ f(0, y, t)\psi(0, y, t) + \int_0^\infty dx \int dy\ f(x, y, 0)\psi(x, y, 0) .$$

On the other hand, computing the total derivative of the function $\psi(\xi_n(t), \eta_n(t), t)$ yields, using the Newton equations (17)

$$\frac{d}{dt}\psi(\xi_n, \eta_n, t) = u(\xi_n, \eta_n, t)\partial_x\psi(\xi_n, \eta_n, t) + E(\xi_n, \eta_n, t)\partial_u\psi + \partial_t\psi .$$

Integrating the above from $t = a_n$ to $t = \infty$ gives

$$\psi(0, \omega_n, a_n) = \int_0^\infty dt \int_0^\infty dx \int dy\ \times$$

$$H(t-a_n)\delta(x-\xi_n(t))\delta(y-\eta_n(t))[u(x, y, t)\partial_x\psi(x, y, t)+E(x, y, t)\partial_u\psi+\partial_t\psi] ,$$

and summing up over all parts gives

$$\int_0^\infty dt \int dy\ \psi(0, y, t) \sum_{n=1}^{N} \delta(t - a_n)\delta(y - r_n) = \tag{19}$$

$$\int f(x, y, t)[u(x, y, t)\partial_x\psi(x, u, t) + E(x, y, t)\partial_y\psi + \partial_t\psi]\ dxdudt$$

Comparing (18) with (19), we see that the measure f satisfies the initial boundary value problem

$$(a)\ \partial_t f+\partial_x[uf]+\partial_y[Ef]=0, \quad (b)\ f(0,y,t)=f^B(y,t)=\sum_{n=1}^{N}\delta(t-a_n)\delta(y-r_n), \tag{20}$$

$$(c)\ f(x,y,0)=0$$

in the weak sense. (Here we start for simplicity with an empty system. A nonzero initial condition can easily be included by assigning parts to the system at time $t = 0$ [4].) The advantage of the kinetic model (20) lies in the fact that the boundary density f^B can be replaced by a continuous function, thus giving a scalable model for a part density. Moments of (20) with respect to the attribute y give the macroscopic equations in section (2.2). If we c.f. postulate that all particles in (20) at a given stage are supported by a single $\delta-$ function of the form

$$f(x,y,t)=\rho(x,t)\delta(y-Y(x,t)) \ ,$$

we obtain the model (2) of the form $\partial_t\rho + \partial_x(v\rho) = 0$ with $v(x,t) = u(x,Y(x,t),t)$. To be compatible with boundary condition (20)(b), we have to choose the macroscopic attribute Y at the boundary such that $Y(0,a_n) = r_n$ holds. This, however is only the most simplistic approach to modeling. The advantage of kinetic models lies in the fact that the relations between attributes and velocities can be much more complex and better adjusted to the real situation.

Models with capacity constraints:
We now turn to a more sophisticated model where we incorporate the fact that the system will have a limited capacity. If the influx into the system exceeds this capacity, bottlenecks will form. This raises the question of policies or service rules, i.e. we have to make a decision which clients / parts to serve first if they cannot all be served simultaneously. The details of the theory outlined below are given in [6]. We consider the case of a production system with a limited capacity where the attribute y denotes the inverse priority of the part. That is, we assume that the total flux at the stage x cannot exceed a certain capacity $\mu(x)$, and we process parts in order of their priority, i.e. parts with a smaller attribute y get processed first until the capacity is reached. The service rule is therefore given by the value we assign the attribute at entry into the system and by the way the

attribute evolves in time. This is modelled by setting the velocity u(x,y,t) in (17) equal to

$$u(x,y,t) = v_0(x)H(b(x,t) - y) \ , \tag{21}$$

where v_0 denotes the free velocity, i.e. the throughput time of stage x, and $b(x,t)$ denotes some cutoff value. This means that parts move with the free velocity v_0 for those attributes $y < b$ and they stop for $y > b$. To enforce the capacity constraint $\mu(x)$ we define the free cumulative flux $G(x,y,t)$ by

$$G(x,y,t) = v_0(x) \int_{-\infty}^{y} f(x,y',t)\ dy' \ ,$$

and the actual flux $F(x,t)$ by

$$F(x,t) = v_0(x) \int H(b(x,t) - y) f(x,y,t)\ dy = G(x, b(x,t), t) \ . \tag{22}$$

To enforce the constraint $F(x,t) \leq \mu(x)$ we have to therefore set $b(x,t)$ equal to $G^{-1}(x,\mu(x),t)$, where G^{-1} denotes the functional inverse of G with respect to its second variable ($G(x,y,t)$ is a monotone function of y and therefore possesses a functional inverse.) Since G is monotone, we can replace the term $H(b(x,t)-y)$ in (21) by $H(G(x,b(x,t),t) - G(x,y,t)) = H(\mu(x) - G(x,y,t))$. This implies for the total flux $F(x,t)$ in (22)

$$F(x,t) = v_0(x) \int H(\mu(x) - G(x,y,t)) f(x,y,t)\ dy = \tag{23}$$

$$\begin{pmatrix} G(x,\infty,t) = v_0(x)\rho(x,t) & \text{for } v_0\rho < \mu \\ \mu(x) & \text{for } v_0\rho > \mu \end{pmatrix} = \min\{\mu(x), v_0(x)\rho(x,t)\} \ .$$

This gives the macroscopic equation postulated in Section (2.3) and in [5]. This is somewhat trivial in the case when parts or clients are processed on a first - come - first - serve basis. It becomes more interesting when the service rule is more complex. Suppose we give each part a deadline or duedate and process parts in order of their remaining time to this duedate. This means we set the time dependent attribute for part number n equal to $\eta_n(t) = d_n - t$, where d_n denotes the duedate. This gives for the boundary density f^B in (20)

$$f^B(y,t) = \sum_{n=1}^{N} \delta(t - a_n)\delta(y - d_n + a_n)$$

and for the kinetic equation

$$\partial_t f + \partial_x[v_0(x)H(\mu(x) - G(x,y,t))f] - \partial_y f = 0, \tag{24}$$

$$G(x,y,t) = v_0 \int_{-\infty}^{y} f(x,y',t)\, dy' .$$

So the attribute y of a part equals the time to due date d_n at the entry time $t = a_n$ and decays continuously, making late parts more important. Now, the conservation law $\partial_t \rho + \partial_x F = 0$ will still hold with the macroscopic flux function F given by (23). However, this equation provides no information about which parts come out first, and whether we met the duedates or not. To obtain this information, we would have to solve the kinetic model (24). The solution of (24) can be quite involved, especially in the case when more than one type of attribute is considered an therefore the attribute variable y is in a higher dimensional space.

The multi - phase model

One way to obtain more information than given by the simple mass conservation law is to consider higher order moments of (24). Integrating (24) against powers of the attribute variable y gives

$$\partial_t m_j + \partial_x F_j - j m_{j-1} = 0,\ j = 0, \ldots 2J - 1 \tag{25}$$

where the moments m_j and the moment fluxes F_j are given by

$$m_j(x,t) = \int y^j f\, dy\,, \tag{26}$$

$$F_j(x,t) = v_0(x) \int y^j \left[H(\mu(x) - G(x,y,t)\right] f(x,y,t)\, dy\ , \tag{27}$$

$$\tag{28}$$

So $m_0 = \rho$ and $F_0 = F$ holds in the previously used notation. This gives J equations for the $2J$ unknowns m_j, F_j. $j = 0, \ldots J - 1$. As often in kinetic theory (see e.g. [17]), the moment system is not closed. Some Ansatz must be made to find additional J relations among these $2J$ data. To express the various unknown fluxes in terms of the moments, we close the expressions by the Ansatz that at each stage x of the process the kinetic density $f(x,y,t)$ is given by a superposition of J $\delta-$ functions and set

$$f(x,y,t) = \sum_{k=1}^{J} n_k(x,t) \delta(y - Y_k(x,t)) . \tag{29}$$

Equation (29) has the following interpretation. By making the Ansatz of a superposition of J concentrations we allow for overtaking of parts, i.e. parts with a higher priority might pass lower priority part. However, we limit the passing process by assuming that no more than J parts can pass each other at the same time, i.e. we consider essentially a traffic model with J lanes. For the case $J = 1$ (single lane traffic, no passing) this reduces to the case when the flux F_0 is given by (23) [6]. This gives rise to what is called multi-phase fluid models, which were originally developed in [14] as an alternative to WKB methods for the Schrödinger equation of quantum mechanics. Using the Ansatz (29), the moments fluxes and acceleration terms in (26) and (27) are given in terms of the concentrations n_k and the macroscopic attributes Y_k in terms of

$$m_j(x,t) = \sum_{k=1}^{J} n_k\, Y_k^j, \tag{30}$$

$$F_j(x,t) = v_0(x) \sum_{k=1}^{J} n_k\, Y_k^j\, Z_k \tag{31}$$

$$Z_k = \max\{0, \min\{1, \frac{\mu(x) - v_0 \sum_{Y_s \neq Y_k} n_s\, H(Y_k - Y_s)}{v_0 \sum_{Y_s = Y_k} n_s}\}\}\,. \tag{32}$$

Inserting these relations into the moment equations (25) yields $2J$ partial differential equations for the $2J$ unknowns $n_1,..,n_J,\ Y_1,..,Y_J$. To illustrate the meaning of the multi-phase model we briefly discuss the case $J = 2$, i.e. we construct a flow with two phases, a high and a low priority phase. In this case, inserting (30) and (31) into the moment equations gives the following set of four equations for n_1, n_2, Y_1, Y_2:

$$\begin{aligned} \partial_t(n_1 Y_1^j + n_2 Y_2^j) + \partial_x v_0(n_1 Z_1 Y_1^j + n_2 Z_2 Y_2^j) \\ = j[n_1 Y_1^{j-1} + n_2 Y_2^{j-1}],\ . \end{aligned} \tag{33}$$

The issue is now the computation of Z_k, $k = 1, 2$. Let us suppose that $Y_1 < Y_2$ to fix the ideas. Then, (32) leads to the following discussion :

$$(\imath)\ \text{if } \mu < n_1 v_0 \quad \text{then} \quad n_1 v_0 Z_1 = \mu \quad \text{and} \quad n_2 v_0 Z_2 = 0\,, \tag{34}$$

$$\begin{aligned} (\imath\imath)\ \text{if } n_1 v_0 < \mu < v_0(n_1 + n_2) \quad &\text{then } n_1 v_0 Z_1 = n_1 v_0 \\ &\text{and} \quad n_2 v_0 Z_2 = \mu - n_1 v_0, \end{aligned} \tag{35}$$

$$\begin{aligned} (\imath\imath\imath)\ \text{if } v_0(n_1 + n_2) < \mu \quad &\text{then} \quad n_1 v_0 Z_1 = n_1 v_0 \\ &\text{and} \quad n_2 v_0 Z_2 = n_2 v_0\,. \end{aligned} \tag{36}$$

Of course, the roles of 1 and 2 must be exchanged in the case $Y_1 > Y_2$. When $Y_1 = Y_2$, then

$$n_1 v_1 Z_1 = \min\{n_1 v_1, \mu \frac{n_1}{n_1 + n_2}\}, \quad n_2 v_2 Z_2 = \min\{n_2 v_2, \mu \frac{n_2}{n_1 + n_2}\}. \tag{37}$$

What formulas (34)-(36) express is very simple. $n_k v_0$ is the 'free flux' of parts in phase k and $v_0(n_1 + n_2)$ is the total 'free flux' (we call 'free fluxes' the fluxes if there would be no flux limitation). In the first case, the flux limitation μ is already below the free flux of parts 1 and therefore, the actual flux of these parts is equal to the flux constraint and parts 2 simply do not move. In the second case, the flux constraint μ is larger than the free flux of parts 1 but below the total free flux. Therefore, the flux constraint does not apply to parts 1 which move with actual flux equal to their free flux. The actual flux constraint which applies to parts 2 is the total flux constraint c diminished by the flux of parts 1 and therefore, parts 2 move under this flux constraint. In the last case, there is no flux constraint at all because the flux constraint is above the total free flux and each parts actually moves according to its own free flux. Clearly, this is consistent with the policy consisting in processing parts with lower attributes first. Again, the role of 1 and 2 must be exchanged in the case $Y_2 < Y_1$. The multiphase model with J phases can be interpreted as a model for a reentrant production system with loops if the phases are linked through the boundary conditions, i.e. for a pull policy we would reduce the phase each time a part runs through the system and set $n_k Y_k Z_k(0,t) = n_{k+1} Y_{k+1} Z_{k+1}(1,t),\ k = 0, .., J-1$ with $n_J Y_J Z_J(0,t)$ equal to the total influx λ.

An example with bottlenecks:
We conclude the discussion of purely deterministic models with an example of a supply chain with two bottlenecks. We solve a 'hot lot' problem, i.e. a system with a steady flow of low priority parts ('cold lots') which is disrupted by a sudden influx of high priority parts ('hot lots'). We consider a chain of 20 stations, all with throughput time =1. So the total minimal throughput time is 20. They all have a capacity of $\mu = 160$ parts per unit time, except for number 5, which has $\mu = 80$ and number 15 which has $\mu = 40$ (two bottlenecks). We consider a constant influx of 'low priority' parts, i.e. with a due date far in the future, of 60 parts per unit time. At time $t = 40$ 'hot lots' (parts with a much closer due date) arrive at a rate of 60 parts per unit time. With these data, the first bottleneck with $\mu = 80$ can accommodate the flow of one of either parts (hot or low) but not both together. The second bottleneck ($\mu = 40$) cannot even accommodate one single flow. Within the low priority lot and the hot lot population the due dates are chosen randomly in a given interval.

The phenomena we expect to see are the following. The low priority lots pass freely through the first bottleneck but start to pile up at station 15. This is the picture until the hot lots arrive at $t = 40$. Once the hot lots arrive, they pass freely through the first bottleneck, but constrict the flow of the low priority lots there. As soon as they reach the second bottleneck, they start to pile up and strangle the low priority flow there completely. Once the hot lots have passed through, the queues start to dissolve. The simulation runs from $t = 0$ to $t = 140$. To verify the two phase model outlined above we have directly simulated the kinetic equation (24) using a particle simulator (see [6] for details). Given the particle solution we generated a two phase approximation to the densities resulting form the particle solution and compared this to the actual solution of the two - phase model (33). So, given the particle solution, we first compute its first four moments $m_0, .., m_3$ and compute a corresponding phase and density according to (30). The corresponding result is compared with the solution of the 2-phase model in Figure 1.6 for different times. The solid and the dashed lines denote the hot and the cold phase of the 2-phase model. The triangles and $\times$'s denote the data points for the corresponding phases extracted from the particle model. (Note, that, numerically, there will always be two phases!). The left panel shows the values of the attributes Y_1 and Y_2, and the right panel shows the densities n_1 and n_2. The densities are plotted on a logarithmic scale. So, for perfect agreement, the $\times$ symbols, the values for the 'cold' phase of the particle model, should be on top of the dashed line, the 'cold' phase of the two phase model. The triangles, the values for the 'hot' phase of the particle model, should be on top of the solid line, the 'hot' phase of the two phase model.

Figure 1.6 shows a reasonable agreement between the particle solution and the 2-phase model. Note, that the two phases coincide for a while to the right of the bottleneck at station 15, meaning that there is only the high priority flux there, since the low priority flux has been cut off completely. At this point both densities (in the right panel) are large since we have parts of both, the hot and the cold phase, accumulating. On the other hand, at the first bottleneck, at station number 5, only the densities of the low priority flux (the symbols $\times$ in the right panel) become large.

1.3.3. *Stochasticity and Diffusion*

The results of Section (1.3.2) pertain to a strictly deterministic system. In essence, we have assumed so far, that the production system works like an automaton, i.e. given a current state of the system we can determine with absolute certainty the further progress of the part traveling through the

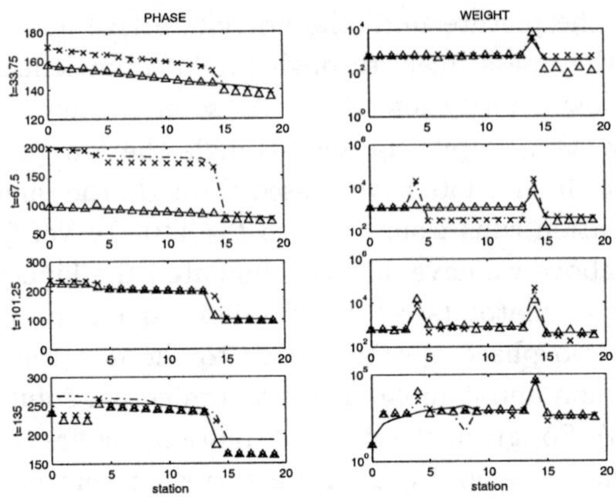

Fig. 1.6. 2-Phase picture, left panel=attributes Y_1, Y_2, right panel=densities, $\times, \triangle$=particles, '-,-.'=2phase model.

system. This approach is insufficient for the following two reasons.

- Production systems are inherently random. Each individual node in the system can incur random breakdowns. The influx in the system will exhibit some statistical variations.
- The system might be too complex to model every single node in detail. Transport parameters, like the form of the velocity $u(x, y, t)$ in (17) might therefore be obtained from observation of the actual physical system over a period of time, measuring WIP's and throughput times, and, rather than fitting one particular function to these data, it will be more accurate to use them in form of statistical distributions.

We therefore introduce the element of randomness into the system. The basic picture for an individual part entering the system at the arrival time a_n with attribute r_n is the following.

Random velocity updates

We consider a production system with M stages. We divide the interval $0 < x < 1$ into M equal subintervals and attribute to each production stage and interval of length $\frac{1}{M}$. Part number n moves through this system with a velocity $\eta_n(t)$ This corresponds to identifying the attributes with velocity and setting $u(x, y, t) = y$ in (17). Each time we enter a new stage we update the velocity η_n, i.e. the estimated throughput time for the next stage, randomly from a given distribution which, in some way, depends on the over all state of the system. In this picture, we would update the velocity randomly each time the part has travelled a distance $\Delta x = \frac{1}{M}$. To obtain a more tractable mathematical model, we replace this picture where we update the velocity not at fixed spatial intervals Δx but, on average, at certain time intervals. So we update the velocity with a certain frequency ω or, in a probabilistic picture, we update the velocity every infinitesimal time step Δt with a probability $\omega \Delta t$. In order to update, on average, each time the part has travelled a distance $\frac{1}{M}$ we have to set $\frac{\eta}{\omega} = \frac{1}{M}$. This corresponds to the concept of a mean free path in gas dynamics [7]. Therefore the advance of a part with position $\xi(t)$ in the infinitesimal time interval Δt is governed by

$$\xi(t + \Delta t) = \xi(t) + \Delta t \eta(t), \quad \eta(t + \Delta t) = (1 - \kappa(\eta(t))\eta(t) + \kappa\gamma(\xi, t) \quad (38)$$

$$d\mathcal{P}\{\kappa(\eta) = k\} = [\Delta t M \eta \delta(k - 1) + (1 - \Delta t M \eta)\delta(k)] \, dk,$$

$$d\mathcal{P}\{\gamma(\xi, t) = z\} = P(\xi, z, t) \, dz,$$

So each Δt we toss a weighted coin and choose $\kappa = 1$ with a probability $\Delta t M \eta$ and $\kappa = 0$ with probability $1 - \Delta t M \eta$. If $\kappa = 1$ holds we update the velocity η randomly from the distribution $P(x, z, t)$ which depends on the state of the production system at time t in some form. If $\kappa = 0$ holds we keep the velocity η at its current level. It should be pointed out here that, at this point we make a form of mean field assumption, namely that the probability distribution P is itself independent of the particle coordinates (ξ, η). P will depend in general of some averaged quantities of the whole ensemble. Taking P independent of (ξ, η) means that we assume that there are many parts in the system, such that the influence of one individual part on the whole ensemble can be neglected. To obtain equations for densities we define the probability distribution

$$f(x, y, t) \, dxdy = d\mathcal{P}\{\xi(t) = x, \eta(t) = y\} \, .$$

At this point we encounter the usual problem defining densities for open systems. The number of parts in the system will not be constant (the part with position ξ does not exist for $t < a$), whereas the probability density $f(x, y, t) \, dxdy$ has to integrate to unity for all time. We remedy this situation by starting the part at some appropriate negative spatial coordinate at time $t = 0$ and move it with a constant velocity for $x < 0$, such that it arrives at the entrance $x = 0$ of the production system precisely at the time a. Thus, we replace (38) by

$$\begin{aligned}(a)\quad \xi(t+\Delta t) &= \xi(t) + \Delta t \eta(t), \\ \eta(t+\Delta t) &= H(-\xi)\eta(t) + H(\xi)[(1-\kappa(\eta(t))\eta(t) + \kappa\gamma] \\ (b)\quad \xi(0) &= -ra, \quad \eta(0) = r\ .\end{aligned} \tag{39}$$

Therefore $\xi(t) = r(t-a)$, $\eta(t) = r$ will hold as long as $t < a$ holds and the part will arrive at $x = 0$ at time $t = a$ with velocity r. Summing up over all possible scenarios we obtain from (39)

$$f(x, y, t + \Delta t) =$$

$$\int \delta(\xi + \Delta t\eta - x)\delta(H(-\xi)\eta + H(\xi)[(1-k)\eta + kz] - y) f(\xi, \eta, t) \times$$

$$d\mathcal{P}\{\kappa(\eta) = k\} d\mathcal{P}\{\gamma(\xi, t) = z\}\ d\xi\eta\ ,$$

which in the limit $\Delta t \to 0$ yields (see [3] for the details)

$$\partial_t f + y\partial_x f = MH(x)[P(x, y, t) \int z f(x, z, t)\ dz - y f(x, y, t)]\ . \tag{40}$$

Assuming that the arrival times a and the initial velocities r are distributed according to a distribution $A(\alpha, \beta)\ d\alpha\beta = d\mathcal{P}\{a = \alpha, r = \beta\}$, we obtain the initial condition

$$f(x, y, 0) = \frac{1}{y} A(-\frac{x}{y}, y)\ .$$

Since f satisfies the pure convection equation $\partial_t f + y\partial_x f = 0$ for $x < 0$, we can replace the above problem by an initial boundary value problem by solving (40) via the method of characteristics for $x < 0$ and obtain the initial boundary value problem

$$\partial_t f + y\partial_x f = M[P(x, y, t) \int z f(x, z, t)\ dz - y f(x, y, t)],\ x > 0, \tag{41}$$

$$y f(0, y, t) = A(t, y)\ .$$

Asymptotics for many stage processes:
To reduce the initial boundary value problem (41) for the kinetic density function f to an equation for the part density ρ in the previous section, we take the zero order moment of (41), and define the quantities

$$\rho(x, t) = \int f(x, y, t)\ dy, \quad F(x, t) = \int y f(x, y, t)\ dy\ . \tag{42}$$

So, $\rho(x,t)\ dx$ denotes the probability of the part being at stage x of the production process at time t, and F is the corresponding flux. Integrating (41) with respect to y gives the mass conservation law

$$\partial_t \rho + \partial_x F = 0, \quad 0 < x < 1, \quad F(0,t) = \lambda(t) = \int A(t,y)\ dy \ , \qquad (43)$$

and the goal is now to express the flux F in terms of the density ρ. This is done by a functional expansion of the kinetic density f in (41) for large values of M, i.e. for a system with many stages, which is the equivalent of the Chapman - Enskog expansion for the solution of the Boltzmann equation of gas dynamics (see c.f. [7]). We formally set

$$f(x,y,t) = \phi(\rho(x,t),y,t) \ .$$

So we try to express the kinetic density f as a 'shape function' dependent on space only through the macroscopic density ρ. Because of the definition (42) the shape function ϕ has to satisfy

$$\int \phi(\rho,y,t)\ dy = \rho, \quad \forall \rho \ . \qquad (44)$$

The macroscopic flux F is then given by

$$F(\rho,x,t) = \int y\phi(\rho(x,t),y,t)\ dy \ .$$

Inserting this Ansatz into the kinetic equation (41), using the macroscopic conservation law (43) gives

$$\partial_t \phi - (\partial_\rho \phi)\partial_x F + \partial_x[y\phi] = M[PF - y\phi] \ , \qquad (45)$$

and we obtain asymptotic expressions for the shape functions ϕ and F by an expansion of (45) for large values of M, i.e. setting $\phi(\rho,y,t) = \phi_0 + \frac{1}{M}\phi_1 + ...$ and $F(\rho,t) = F_0 + \frac{1}{M}F_1 + ...$ For the zero'th order term of this expansion we obtain the equation

$$PF_0 - y\phi_0 = 0, \quad F_0 = \int y\phi_0\ dy \ ,$$

which implies that ϕ_0 is given by the probability density P divided by y and multiplied by an arbitrary function of x and t. We denote this arbitrary function by ρv and obtain the macroscopic velocity $v(x,t)$ from the normalization condition (44), giving

$$\phi_0(\rho,x,y,t) = \rho v(x,t)\frac{P(x,y,t)}{y}, \quad \frac{1}{v(x,t)} = \int \frac{P(x,y,t)}{y}\ dy, \qquad (46)$$

$$F_0(\rho, x, t) = \rho v(x, t)$$

This gives in zero'th order the heuristic model (9) with the macroscopic velocity v computed as the reciprocal value of the reciprocal microscopic velocity. For a part moving with velocity y the random throughput time $\tau(x, t)$ for the station at stage x, occupying an interval of length $\frac{1}{M}$, is given by $\tau = \frac{1}{My}$. Therefore the macroscopic velocity $v(x, t)$ is related to the random throughput time as

$$v(x,t) = \frac{1}{ME[\tau(x,t)]}, \tag{47}$$

where $E[\tau]$ denotes the expectation of τ under the probability distribution P. This fact is important if one wishes to generate the distribution P from observed data of an actual production system. Going to the next term in the expansion will give the diffusive term. The balance of order $O(\frac{1}{M})$ terms in (45) reads

$$\partial_t \phi_0 - (\partial_\rho \phi_0)\partial_x F_0 + \partial_x[y\phi_0] = PF_1 - y\phi_1$$

or using (46)

$$\rho \partial_t[\frac{vP}{y}] - \frac{vP}{y}\partial_x(v\rho) + \partial_x[\rho v P] = PF_1 - y\phi_1$$

The shape F_1 is again computed from the normalization condition (45), which in first order reads $\int \phi_1(\rho, x, y, t)\, dy = 0\ \forall \rho$. This gives

$$\rho \partial_t[v \int \frac{P}{y^2}\, dy] - v\partial_x(v\rho) \int \frac{P}{y^2}\, dy + \partial_x[\rho v \int \frac{P}{y}\, dy] = F_1 \int \frac{P}{y}\, dy$$

or

$$F_1 = v\rho\partial_t[v \int \frac{P}{y^2}\, dy] - v^2\partial_x(v\rho) \int \frac{P}{y^2}\, dy + v\partial_x\rho = -D\partial_x\rho + vR\rho$$

with the diffusion coefficient D and the first order correction R of the velocity being given in terms of the variation coefficient $V[\frac{1}{y}]$ of the throughput time $\frac{1}{y}$, i.e. the variance of this quantity scaled by the square of the mean. D and R can be expressed as

$$(a)\ D(x,t) = vV[\frac{1}{y}], \quad V[\frac{1}{y}] = v^2 \int \frac{1}{y^2} P\, dy - 1 = \frac{E[\frac{1}{y^2}] - E[\frac{1}{y}]^2}{E[\frac{1}{y}]^2}, \tag{48}$$

$$(b)\ R = \partial_t(\frac{V[\frac{1}{y}] + 1}{v}) - (V[\frac{1}{y}] + 1)\frac{1}{v}\partial_x v = (\partial_t + v\partial_x)\frac{V[\frac{1}{y}] + 1}{v} - \partial_x V[\frac{1}{y}]$$

$V[\frac{1}{y}]$ is a dimensionless measure of the stochasticity of the system. $V = 0$ holds for the degenerate case if the probability distribution P is a $\delta-$ function concentrated on the mean. It is important to notice that the diffusion coefficient will always be positive. Thus, the flux in the continuity equation (43) is given up to order $\frac{1}{M^2}$ terms as

$$F = \rho v(1 + \frac{R}{M}) - \frac{1}{M} v V[\frac{1}{y}] \partial_x \rho \ . \tag{49}$$

Note that, because of (47) the macroscopic velocity v, and therefore the whole flux F, will be of order $O(\frac{1}{M})$. The reason for this is that we measure time in units corresponding to the throughput time of one individual node in the supply chain. So the whole system will therefore evolve very slowly on this time scale. Rescaling time velocity and fluxes via

$$t \to tM, \quad \partial_t \to \frac{\partial_t}{M}, \quad v \to \frac{v}{M}, \quad F \to \frac{F}{M}, \quad \lambda(t) \to \frac{1}{M}\lambda(\frac{t}{M})$$

would yield the same initial value problem for the conservation law (43) with the same flux function F as in (49), but with a macroscopic velocity $v(x,t)$, according to (47) which is now of order $O(1)$.

So far, $\rho(x,t)$ has denoted the probability distribution that a certain part is at stage x at time t and $\lambda(t)$ was the probability distribution that the part arrives in the system at time t. Under the mean field assumption that the density P is independent of a single part, all parts in the system will obey the same equation. Therefore the probability ρ can be identified with the part density ρ up to a constant factor and the probability density λ can be identified with the start rate (the influx into the system) up to the same factor. In zero'th order (for $M = \infty$) the flux function (49) reduces to the one given in (2) with $\frac{1}{v}$ given by the expectation of the local throughput time τ. In first order, including the $O(\frac{1}{M})$ terms in (49) we obtain the diffusion model (9) with the diffusion coefficient given in terms of the variation coefficient of the statistics. (The term $\frac{R}{M}$ can be neglected since it will always be dominated by v.) However, the above analysis gives an indication of how to match the transport coefficients v and D to a given production system in a more sophisticated way than by just fitting parameters.

The usage of the above analysis is the following. To obtain a simple model of an actual supply chain we wish to determine transport coefficients for the conservation law (43) from observations of the system for a certain period of time. To this end we break down the process into M stages and observe the time each part has arrived at each stage for a large number of parts, i.e. we record the numbers t_n^m, $n = 1,..,N$, $m = 1,..,M+1$ for $N >> 1$, where t_n^{M+1} denotes the exit time. Then we proceed as follows.

- From this we compute the throughput times $\tau_n^m = t_n^{m+1} - t_n^m$, $n = 1,..,N$, $m = 1,..,M$, i.e. the time it took part number n to complete stage number m of the process.
- We also record a certain integral quantity (like the WIP) of the whole system, computing a state variable $S_m(t_n^m)$ for each stage.
- Then we fit a probability distribution $T_m(\tau, S_m(t_n^m))$ to the throughput times τ_n^m.
- After making everything continuous, we obtain a probability distribution $T(x, \tau, S)\ d\tau$ for the throughput time through the stage x, given a state S of the system. This probability density is related to the probability density for the velocity in the above derivation via $P(x, y, S) = \frac{1}{My^2} T(x, \frac{1}{My}, S)$.
- Then we compute the transport coefficients v, D and R from (47) and (48), which are now dependent on the state variable $S(x,t)$ as well.

After completing this process we have a relatively simple model of the form

$$(a)\ \partial_t \rho + \partial_x F = 0, \quad 0 < x < 1, \quad F(0,t) = \lambda(t)\ , \tag{50}$$

$$(b)\ F = [1 + \frac{1}{M} R(x,t,S)]\rho v(x,t,S) - \frac{1}{M} vV[\frac{1}{y}](x,t,S)\partial_x \rho$$

for the supply chain. It remains to define the state variable $S_m(t_n^m)$, or in the continuous version $S(x,t)$. This is a question of how to interpret given experimental data and what integral state of the system to record in the experiment. For a linear supply chain without any reentrant processes it will be reasonable to take $S_m(t_n^m)$ equal to the number of parts in stage m at time t_n^m, or to take $S(x,t) = \rho(x,t)$. For a more complex system with reentrant processes, it will be more appropriate to take S equal to the total work in progress, i.e. $S(x,t) = S(t) = \int \rho(x,t)\ dx$. If the supply chain contains reentrant processes which are solely governed by pull policies, then the throughput times will depend only on the load downstream and we will set $S(x,t) = \int_x^1 \rho(x',t)\ dx'$.

A numerical comparison

We conclude this section with a numerical example. We consider a reentrant production system, for which we prescribe a uniform spatial velocity distribution between a minimum and a maximum local throughput time $\tau(x)$. We simulate the system via the zeroth order model (2) taking v as the expectation. This is referred to as the queuing model. Then we include the random updates according to (39) in a particle based simulation (see [3] for details). Figure 1.7 shows the outflux of the system for the deterministic model (the dashed line) and a time averaged version of the randomized velocity updates (the solid line). We see the effect of randomness lowering the initial outflux significantly.

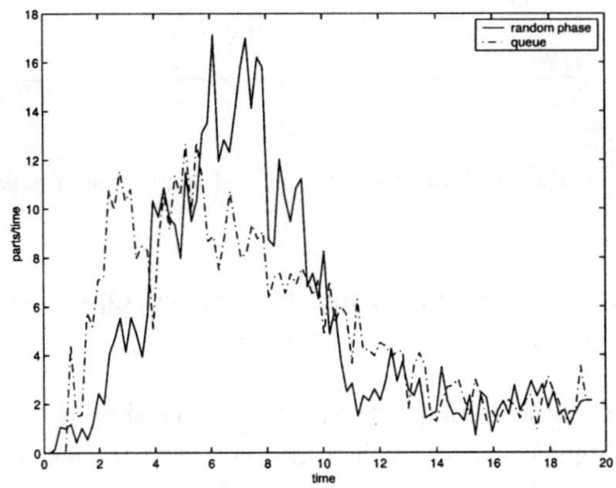

Fig. 1.7. Outflux (Random phase and queue) N=400 lots, 20 ensembles.

Figure 1.8 compares the outflux of the time averaged particle model with the conservation law when diffusion is added (i.e. all the terms in the flux function in (50) are considered.

1.4. Conclusions

We have shown how the heuristic models for production flows (Eqns 2, 9, 14) arise from different closure models of expansions of the Boltzmann equation (Eq.15). These first principle models can also be extended beyond simple heuristics to generate dynamical models for production flows with dispatch rules based on the evolution of an attribute (e.g. the duedate of a particular product). The present review is a snapshot of the state of a larger research project to develop the fundamental macroscopic dynamical equations for production flows based on kinetic theory of the motion of individual production lots, and their interactions based on production rules

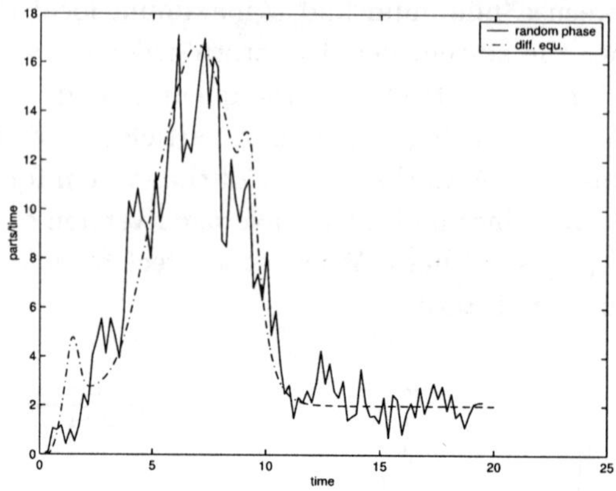

Fig. 1.8. Outflux for the random phase model and the diffusion equation, N=400 lots, 20 ensembles.

and competition for machine time and floor space. Open problems currently under investigation are:

- A macroscopic transport equation reflecting the fact that the capacity along the production line varies stochastically.
- Detailed studies comparing large scale discrete event simulations with the different macroscopic models. This involves in particular studies of time dependent flows - transient or stochastic. Open questions involve the errors generated by treating a fundamentally discrete flow as a continuum flow, the errors resulting from quasi-static and diffusive models for flows changing on a fast timescale and the ability to extract the diffusion coefficient and the state equation from discrete event data.
- The kinetic models for the flow evolution with attributes will allow us to make connections to scheduling algorithms [21]. In discrete time and for discrete production many of these algorithms are NP-hard. It remains to be seen whether a continuum approximation will lead to simpler optimization problems that are still useful as scheduling rules.

Based on the dichotomies discussed in the introduction between industrial production networks and biomolecular networks we also think that the dialogue between the two fields should be continued and extended. As a minimum, thinking about the differences in the two fields sharpens the awareness of the essential features of industrial production or biological production, respectively. However, beyond introspection, we are also quite optimistic that techniques and insights can actually be transferred between these fields.

Acknowledgements

This work was supported in parts by a grant from Intel Corporation and by NSF grant DMS 0204543. We thank Karl Kempf, Erjen Lefeber and Dan Rivera for many insightful discussions and Tae-Chang Jo and Roel van den Berg for computational assistance.

References

1. D. Armbruster, C. Ringhofer, T-J. Jo, Continuous models for production flows, in: Proceedings of the 2004 American Control Conference, Boston, pp 4589 - 4594, (2004).
2. Dieter Armbruster, Daniel Marthaler, Christian Ringhofer, Karl Kempf, Tae-Chang Jo: A continuum model for a re-entrant factory, in revision for Operations research 38 pages 9/2003
3. D. Armbruster, C. Ringhofer: "Thermalized kinetic and fluid models for re-entrant supply chains", SIAM J. Multiscale Modeling and Simulation, **3**(4), pp 782 - 800, (2005).
4. D. Armbruster, D. Marthaler, C. Ringhofer: Kinetic and Fluid Model Hierarchies for Supply Chains, *SIAM J. on Multiscale Modeling* 2, pp.43-61 (2004).
5. D. Armbruster, P. Degond, C. Ringhofer: A Model for the Dynamics of large Queuing Networks and Supply Chains, submitted, 2004
6. D. Armbruster, P. Degond, C. Ringhofer: "Kinetic and fluid models for supply chains supporting policy attributes submitted, (2004).
7. S. G. Brush: Kinetic Theory, *Pergamon Press* (1972).
8. Jakob Asmundsson, Reha Uzsoy, and Ronald L. Rardin: Compact nonlinear capacity models for supply chains: Methodology, preprint, 2002, Purdue University
9. Bartholdi, J.J. III, D. D. Eisenstein, A production line that balances itself, *Operations Research*, **44:1** (1996) 21–34.
10. H. Baumgaertel, S. Brueckner, V. Parunak, R. Vanderbok, and J. Wilke. "Agent Models of Supply Network Dynamics." in Terry Harrison et al (Eds), The Practice of Supply Chain Management, Kluver, 2003
11. C. Cercignani: The Boltzmann equation and its applications, *Applied Mathematical Sciences, vol. 67, Springer Verlag* (1988).
12. Chi reference manual at http://se.wtb.tue.nl/documentation/
13. J. G. Dai, J.H. Vande Vate, The stability of two-station multitype fluid networks, Operations Research, **48**(5), 721-744 (2000)
14. S. Jin, X. Li, *Multi-phase Computations of the Semiclassical Limit of the*

Schrodinger Equation and Related Problems: Whitham vs. Wigner, Physica D **182**, pp. 46-85 (2003).
15. K. H. Karlsen, N. H. Risebro, J. D. Towers, *L^1 stability for entropy solutions of nonlinear degenerate parabolic convection-diffusion equations with discontinuous coefficients*, Skr., K. Nor. Vidensk. Selsk. **3** pp. 1–49 (2003).
16. A. Klar, R. Wegener, *Enskog-like kinetic models for vehicular traffic*, J. Stat. Phys., (1997), pp. 91–114.
17. C. D. Levermore, *Moment closure hierarchies for kinetic theories*, J. Stat. Phys., **83** (1996), pp. 1021–1065.
18. M.J. Lighthill, G.B. Whitham, On kinematic waves II. A theory of traffic flow on long crowded roads, Proceedings of the Royal Society, Series A, **229**, 317-345 (1955)
19. U.S. Karmarkar: Capacity loading and release planning in Work-in-Progess (WIP) and Lead-times, J. Mfg. Oper.Mgt., 2, 105 - 123 (1989)
20. B. Hess, A.S. Mikhailov, Science **264**, 223 (1994)
21. Michael Pinedo, *Scheduling*, Prentice Hall, 1995
22. Bart Rem, Dieter Armbruster, Control and Synchronization in Switched Arrival Systems, Chaos **13** (1), 128-137 (2003)
23. Collaborative Research Centre 637 "Autonomous Cooperating Logistics Processes: A Paradigm Shift and its Limitations" Universität Bremen http://www.sfb637.uni-bremen.de

CHAPTER 2

SUPPLY AND PRODUCTION NETWORKS: FROM THE BULLWHIP EFFECT TO BUSINESS CYCLES

Dirk Helbing, Stefan Lämmer

Institute for Transport & Economics, Dresden University of Technology
Andreas-Schubert-Str. 23, D-01062 Dresden, Germany
E-mail: helbing@trafficforum.org, traffic@stefanlaemmer.de

Network theory is rapidly changing our understanding of complex systems, but the relevance of topological features for the dynamic behavior of metabolic networks, food webs, production systems, information networks, or cascade failures of power grids remains to be explored. Based on a simple model of supply networks, we offer an interpretation of instabilities and oscillations observed in biological, ecological, economic, and engineering systems.

We find that most supply networks display damped oscillations, even when their units – and linear chains of these units – behave in a non-oscillatory way. Moreover, networks of damped oscillators tend to produce growing oscillations. This surprising behavior offers, for example, a new interpretation of business cycles and of oscillating or pulsating processes. The network structure of material flows itself turns out to be a source of instability, and cyclical variations are an inherent feature of decentralized adjustments. In particular, we show how to treat production and supply networks as transport problems governed by balance equations and equations for the adaptation of production speeds.

The stability and dynamic behavior of supply networks is investigated for different topologies, including sequential supply chains, "supply circles", "supply ladders", and "supply hierarchies". Moreover, analytical conditions for absolute and convective instabilities are derived. The empirically observed bullwhip effect in supply chains is explained as a form of convective instability based on resonance effects. An application of this theory to the optimization of production networks has large optimization potentials.

2.1. Introduction

Supply chain management is a major subject in economics, as it significantly influences the efficiency of production processes. Many related studies focus on subjects such as optimum buffer sizes and stock levels.[1,2,3,4] However, the optimal structure of supply and distribution networks is also an important topic,[5,6] which connects this scientific field with the statistical physics of networks.[7,8,9,10,11,12,13,14] Problems like the adaptivity and robustness of supply networks have not yet been well covered and are certainly an important field for future research.

In this chapter, we will focus on the dynamical properties and linear stability of supply networks in dependence of the network topology or, in other words, the supply matrix (input matrix). Presently, there are only a few results on this subject, since the response of supply networks to varying demand is not a trivial problem, as we will see. Some "fluid models" have, however, been proposed to study this subject. Daganzo,[15] for example, applies the method of cumulative counts[16] and a variant of the Lighthill-Whitham traffic model to study the so-called bullwhip effect.[17,18,19,20,21,22,23,24,25,26,27,28] This effect has, for example, been reported for beer distribution,[29,30] but similar dynamical effects are also known for other distribution or transportation chains. It describes increasing oscillations in the delivery rates and stock levels from one supplier to the next upstream supplier delivering to him.

Similar models are studied by Armbruster *et al.*[31,32,33,34] Most publications, however, do not investigate the impact of the network topology, but focus on sequential supply chains or re-entrant problems.[34,35] An exception is the transfer function approach by Dejonckheere *et al.*[24] as well as Disney and Towill.[26] Ponzi, Yasutomi, and Kaneko[36] have coupled a supply chain model with a model of price dynamics, while Witt and Sun[37] have suggested to model business cycles analogously to stop-and-go traffic. Similar suggestions have been made by Daganzo[15] and Helbing.[38] We should also note the relationship with driven many-particle models.[39] This is the basis of event-driven simulations of production systems, which are often used in logistics software.

In a previous publication, one of the authors has proposed a rather general model for supply networks, and has connected it to queueing theory and macroeconomics.[38] That paper also presents numerical studies of the bullwhip effect in sequential supply chains and the effect of heterogeneous production units on the dynamics of supply networks. Moreover, the dependence of the maximum oscillation amplitude on the model parameters has been numerically studied, in particular the phase transition from stable to bullwhip behavior. The underlying simulation models are of non-linear nature,[40] so that phenomena such as wave selection and synchronization have been observed.[38] Finally, a large variety of management strategies including the effect of forcasting inventories has been studied together with Nagatani[38,41]. The latter has also investigated the dynamical effect of multiple production lines.[42]

In the following review, we will mainly focus on a linearized model of supply networks, which could be viewed as a dynamical version[43] of Leontief's classical input-output model.[44] It is expected to be valid close to the stationary state of the supply system or, in other words, close to the (commonly assumed) economic equilibrium. This model allows to study many properties of supply networks in an analytical way, in particular the effects of delayed adaptation of production rates or of a forecasting of inventories. In this way, we will be able to understand many numerically and empirically observed effects, such as resonance effects and the stability properties of supply networks in dependence of the network topology.

Our review is structured as follows: Section 2.2 introduces our model of supply networks and linearizes it around the stationary state. Section 2.2.2 discusses the bullwhip effect for sequential supply chains. Surprisingly, the amplification of amplitudes does not require unstable eigenvalues, but it is based on a resonance effect. Sec. 2.3.1 presents general methods of solution for arbitrary topologies of supply networks. It reveals some useful properties of the eigenvalues characterizing the stability of supply systems. Interestingly enough, it turns out that all supply networks can be mapped on formulas for linear (serial) supply chains. Sec. 2.3.2 will investigate, how the dynamic behavior of supply networks depends on their network structure, while Sec. 2.4 gives an interpretation of business cycles based on material flow networks. Sec. 2.6 finally summarizes our results and points to some future research directions.

2.2. Input-Output Model of Supply Networks

Our production model assumes u production units or suppliers $j \in \{1, \ldots, u\}$ which deliver d_{ij} products of kind $i \in \{1, \ldots, p\}$ per production cycle to other suppliers and consume c_{kj} goods of kind k per production cycle. The coefficients c_{kj} and d_{ij} are determined by the respective production process, and the number of production cycles per unit time (e.g. per day) is given by the production speed $Q_j(t)$. That is, supplier j requires an average time interval of $1/Q_j(t)$ to produce and deliver d_{ij} units of good i. The temporal change in the number $N_i(t)$ of goods of kind i available in the system is given by the difference between the inflow

$$Q_i^{\text{in}}(t) = \sum_{j=0}^{u} d_{ij} Q_j(t) \tag{1}$$

and the outflow

$$Q_i^{\text{out}}(t) = \sum_{i=1}^{u+1} c_{ij} Q_j(t) \,. \tag{2}$$

In other words, it is determined by the overall production rates $d_{ij}Q_j(t)$ of all suppliers j minus their overall consumption rates $c_{ij}Q_j(t)$:

$$\frac{dN_i}{dt} = Q_i^{\text{in}}(t) - Q_i^{\text{out}}(t) = \underbrace{\sum_{j=1}^{u} d_{ij} Q_j(t)}_{\text{supply}} - \underbrace{\left[\sum_{j=1}^{u} c_{ij} Q_j(t) + Y_i(t) \right]}_{\text{demand}} \,. \tag{3}$$

In this dynamic variant of Leontief's classical input-output model,[44] the quantity

$$Y_i(t) = \underbrace{c_{i,u+1} Q_{u+1}(t)}_{\text{consumption and losses}} - \underbrace{d_{i0} Q_0(t)}_{\text{inflow of resources}} \tag{4}$$

comprises the consumption rate of goods i, losses, and waste (the "export" of material), minus the inflows into the considered system (the "imports"). In the following, we will assume that the quantities are measured in a way that $0 \le c_{ij}, d_{ij} \le 1$ (for $1 \le i \le p$, $1 \le j \le u$) and that the "normalization conditions"

$$d_{i0} = 1 - \sum_{j=1}^{u} d_{ij} \ge 0 \,, \qquad c_{i,u+1} = 1 - \sum_{j=1}^{u} c_{ij} \ge 0 \tag{5}$$

are fulfilled. Equations (3) can then be interpreted as conservation equations for the flows of goods.

2.2.1. *Adaptation of Production Speeds*

In addition, we have formulated an equation for the production or delivery rate $Q_j(t)$. Changes in the consumption rate $Y_i(t)$ sooner or later require an adaptation of $Q_j(t)$. For the temporal change dQ_j/dt of the delivery rate we will assume:

$$\frac{dQ_j}{dt} = F_j(\{N_i(t)\}, \{dN_i/dt\}, \{Q_l(t)\}) . \tag{6}$$

Herein, the curly brackets indicate that the so-called management or control function $F_j(\dots)$ may depend on all inventories $N_i(t)$ with $i \in \{1, \dots, p\}$, their derivatives dN_i/dt, and/or all production speeds $Q_l(t)$ with $l \in \{1, \dots, u\}$. Some reasonable specifications will be discussed later on.

2.2.2. *Modelling Sequential Supply Chains*

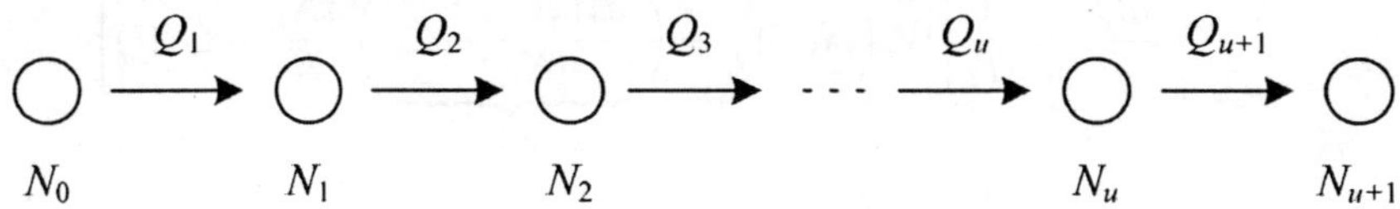

Fig. 2.1. Illustration of the linear supply chain treated in this chapter, including the key variables of the model. Circles represent different suppliers i, N_i their respective stock levels, and Q_i the delivery rate to supplier i or the production speed of this supplier. $i = 0$ corresponds to the resource sector generating the basic products and $i = u + 1$ to the consumers.

For simplicity, let us first investigate a model of sequential supply chains (see Fig. 2.1), which corresponds to $d_{ij} = \delta_{ij}$ and $c_{ij} = \delta_{i+1,j}$. In other words, the products i are directly associated with producers j and we have an input matrix of the form

$$\mathbf{C} = \begin{pmatrix} 0\,1\,0\,0\,0 \\ 0\,0\,1\,0\,0 \\ 0\,0\,0\,1\,0 \\ 0\,0\,0\,0\,1 \\ 0\,0\,0\,0\,0 \end{pmatrix} . \tag{7}$$

This implies $Q_i^{\text{in}}(t) = Q_i(t) = Q_{i-1}^{\text{out}}(t)$ and $Q_i^{\text{out}}(t) = Q_{i+1}(t) = Q_{i+1}^{\text{in}}(t)$, so that the inventory of goods of kind i changes in time t according to

$$\frac{dN_i}{dt} = Q_i^{\text{in}}(t) - Q_i^{\text{out}}(t) = Q_i(t) - Q_{i+1}(t) . \tag{8}$$

The assumed model consists of a series of u suppliers i, which receive products of kind $i-1$ from the next "upstream" supplier $i-1$ and generate products of kind i for the next "downstream" supplier $i+1$ at a rate $Q_i(t)$.[45,31]

The final products are delivered at the rate $Q_u(t)$ and removed from the system with the consumption rate $Y_u(t) = Q_{u+1}(t)$.

In the following, we give some possible specifications of equation (6) for the production rates:

- The delivery rate $Q_i(t)$ may, for example, be adapted to a certain desired rate $\hat{W}_i\,(N_i, dN_i/dt)$ according to the equation
$$\frac{dQ_i}{dt} = \frac{1}{T_i}\left[\hat{W}_i\left(N_i, \frac{dN_i}{dt}\right) - Q_i(t)\right] , \tag{9}$$
where T_i denotes the relaxation time.
- A special case of this is the equation
$$\begin{aligned}\frac{dQ_i}{dt} &= \frac{1}{T_i}\left[W_i\Big(N_i(t)\Big) + \underbrace{\left(\frac{dW_i(N_i)}{dN_i}\,\Delta t\right)}_{=\beta_i}\frac{dN_i}{dt} - Q_i(t)\right] \\ &\approx \frac{1}{T_i}\left[W_i\Big(N_i(t+\Delta t)\Big) - Q_i(t)\right] . \end{aligned}\tag{10}$$
Herein, the desired production speed $W_i(N_i)$ is monotonously falling with increasing inventories N_i, which are forecasted over a time period Δt.
- A further specification is given by the equation
$$\frac{dQ_i}{dt} = \frac{1}{T_i}\left\{\frac{N_i^0 - N_i(t)}{\tau_i} - \beta_i\frac{dN_i}{dt} + \epsilon_i[Q_i^0 - Q_i(t)]\right\} , \tag{11}$$
which assumes that the production rate is controlled in a way that tries to reach some optimal inventory N_i^0 and production rate Q_i^0, and attempts to miminize changes dN_i/dt in the inventory to reach a constant work in progress (CONWIP strategy). Note that at least one of the parameters T_i, τ_i, β_i and ϵ_i could be dropped.
- In other cases, it can be reasonable to work with a model focussing on relative changes in the variables:
$$\frac{dQ_i/dt}{Q_i} = \hat{\nu}_i\left(\frac{N_i^0}{N_i(t)} - 1\right) - \hat{\mu}_i\frac{dN_i/dt}{N_i(t)} + \hat{\epsilon}_i\left(\frac{Q_i^0}{Q_i(t)} - 1\right) . \tag{12}$$
This model assumes large production rates when the inventory is low and prevents that $Q_i(t)$ can fall below zero. Apart from this, $N_i(t)$ and $Q_i(t)$ are adjusted to some values N_i^0 and Q_i^0, which are desireable from a production perspective (i.e. in order to cope with the breakdown of machines, variations in the consumption rate, etc.) Moreover, the control strategy (12) counteracts temporal

changes dN_i/dt in the inventory. For comparison with the previous strategies, we set $T_i = 1/Q_i^0$, $\tau_i = N_i^0/\hat{\nu}_i$, $\beta_i = \hat{\mu}_i/N_i^0$, and $\epsilon_i = \hat{\epsilon}_i T_i$.

The above control strategies appear to be appropriate to keep the inventories $N_i(t)$ stationary, to maintain a certain optimal inventory N_i^0 (in order to cope with stochastic variations due to machine breakdowns etc.), and to operate with the equilibrium production rates $Q_i^0 = Y_u^0$, where Y_u^0 denotes the average consumption rate. However, the consumption rate is typically subject to perturbations, which may cause a bullwhip effect, i.e. growing variations in the stock levels and deliveries of upstream suppliers (see Sec. 2.2.4).

2.2.3. *More Detailed Derivation of the Production Dynamics (with Dieter Armbruster)*

Let us assume a sequential supply chain and let $Q_i(t)$ be the influx in the last time period $T = 24$ hours from the supplier $i-1$ to producer i. This influx is equal to the release $R_{i-1}(t)$ over the last 24 hours:

$$Q_i(t) = R_{i-1}(t) . \tag{13}$$

Based on the current state (the inventory and fluxes at time t and before), a price $P_i(t)$ is set according to some pricing policy and instantaneously communicated downstream to the customer $i+1$. Then, based on price, fluxes, and inventory, the order quantity $D_{i+1}(t)$ is decided at the end of day t according to some order policy W_{i+1}:

$$D_{i+1}(t) = W_{i+1}\Big(N_{i+1}(t), P_i(t), \dots\Big) . \tag{14}$$

It determines the upstream release $R_i(t+T)$ on the next day $t+T = t+1$:

$$R_i(t+T) = R_i(t+1) = D_{i+1}(t) . \tag{15}$$

Altogether, this implies

$$Q_i(t+T) = Q_i(t+1) = R_{i-1}(t+1) = D_i(t) = W_i\Big(N_i(t), \dots\Big) \tag{16}$$

and the following equation for the change in the production rate:

$$\frac{dQ_i}{dt} \approx \frac{\Delta Q_i(t)}{T} = \frac{1}{T}[Q_i(t+T) - Q_i(t)] = \frac{1}{T}\Big[W_i\Big(N_i(t), \dots\Big) - Q_i(t)\Big] . \tag{17}$$

This is consistent with Eq. (9). Moreover, we obtain the usual balance equation for the change of inventory in time:

$$\frac{dN_i}{dt} \approx \frac{\Delta N_i}{T} = \frac{N_i(t+T) - N_i(t)}{T} = N_i(t+1) - N_i(t) = R_{i-1}(t) - R_i(t)$$
$$= Q_i(t) - Q_{i+1}(t) \,. \tag{18}$$

2.2.4. *Dynamic Solution and Resonance Effects (with Tadeusz Płatkowski)*

Let us now calculate the dynamic solution of the sequential supply chain model for cases where the values N_i^0 and Q_i^0 correspond to the stationary state of the production system.[6,39] Then, the linearized model equations for the control approaches (9) to (12) are exactly the same. Representing the deviation of the inventory from the stationary one by $n_i(t) = N_i(t) - N_i^0$ and the deviation of the delivery rate by $q_i(t) = Q_i(t) - Q_i^0$, they read

$$\frac{dq_i}{dt} = \frac{1}{T_i}\left[-\frac{n_i(t)}{\tau_i} - \beta_i \frac{dn_i}{dt} - \epsilon_i q_i(t)\right] . \tag{19}$$

Moreover, the linearized equations for the inventories are given by

$$\frac{dn_i}{dt} = q_i(t) - q_{i+1}(t) \,. \tag{20}$$

Deriving Eq. (19) with respect to t and inserting Eq. (20) results in the following set of second-order differential equations:

$$\frac{d^2 q_i}{dt^2} + \underbrace{\frac{(\beta_i + \epsilon_i)}{T_i}}_{=2\gamma_i} \frac{dq_i}{dt} + \underbrace{\frac{1}{T_i \tau_i}}_{={\omega_i}^2} q_i(t) = \underbrace{\frac{1}{T_i}\left[\frac{q_{i+1}(t)}{\tau_i} + \beta_i \frac{dq_{i+1}}{dt}\right]}_{=f_i(t)} . \tag{21}$$

This corresponds to the differential equation of a damped harmonic oscillator with damping constant γ, eigenfrequency ω, and driving term $f_i(t)$. The eigenvalues of these equations are

$$\lambda_{1,2} = -\gamma_i \pm \sqrt{{\gamma_i}^2 - {\omega_i}^2} = -\frac{1}{2T_i}\left[(\beta_i + \epsilon_i) \mp \sqrt{(\beta_i + \epsilon_i)^2 - 4T_i/\tau_i}\right] . \tag{22}$$

For $(\beta_i + \epsilon_i) > 0$ their real parts are always negative, corresponding to a stable behavior in time. Nevertheless, we will find a *convective* instability, i.e. the oscillation amplitude can grow from one supplier to the next one upstream.

Assuming periodic oscillations of the form $f_i(t) = f_i^0 \cos(\alpha t)$, the general solution of Eq. (21) is of the form

$$q_i(t) = f_i^0 F_i \cos(\alpha t + \varphi_i) + D_i^0 \mathrm{e}^{-\gamma_i t} \cos(\Omega_i t + \theta_i) \,, \tag{23}$$

where the parameters D_i^0 and θ_i depend on the initial conditions. The other parameters are given by

$$\tan\varphi_i = \frac{2\gamma_i\alpha}{\alpha^2 - \omega_i{}^2} = \frac{\alpha(\beta_i + \epsilon_i)}{\alpha^2 T_i - 1/\tau_i}, \tag{24}$$

$$\Omega_i = \sqrt{\omega_i{}^2 - \gamma_i{}^2} = \frac{1}{T_i}\sqrt{1/\tau_i{}^2 - (\beta_i + \epsilon_i)^2/4}, \tag{25}$$

and

$$F_i = \frac{1}{\sqrt{(\alpha^2 - \omega_i{}^2)^2 + 4\gamma_i{}^2\alpha^2}} = \frac{T_i}{\sqrt{(\alpha^2 T_i - 1/\tau_i)^2 + \alpha^2(\beta_i + \epsilon_i)^2}}. \tag{26}$$

The dependence on the eigenfrequency ω_i is important for understanding the occuring resonance effect, which is particularly likely to appear, if the oscillation frequency α of the consumption rate is close to one of the resonance frequencies ω_i. After a transient time much longer than $1/\gamma_i$ we find

$$q_i(t) = f_i^0 F_i \cos(\alpha t + \varphi_i). \tag{27}$$

Equations (21) and (27) imply

$$f_{i-1}(t) = \frac{1}{T_i}\left[\frac{q_i(t)}{\tau_i} + \beta_i \frac{dq_i}{dt}\right] = f_{i-1}^0 \cos(\alpha t + \varphi_i + \delta_i) \tag{28}$$

with

$$\tan\delta_i = \alpha\beta_i\tau_i \qquad \text{and} \qquad f_{i-1}^0 = f_i^0 \frac{F_i}{T_i}\sqrt{(1/\tau_i)^2 + (\alpha\beta_i)^2}. \tag{29}$$

Therefore, the set of equations (21) can be solved successively, starting with $i = u$ and progressing to lower values of i.

2.2.5. *The Bullwhip Effect*

The oscillation amplitude increases from one supplier to the next upstream one, if

$$\frac{f_{i-1}^0}{f_i^0} = \left\{1 + \frac{\alpha^2[\epsilon_i(\epsilon_i + 2\beta_i) - 2T_i/\tau_i] + \alpha^4 T_i{}^2}{(1/\tau_i)^2 + (\alpha\beta_i)^2}\right\}^{-1/2} > 1. \tag{30}$$

One can see that this resonance effect can occur for $0 < \alpha^2 < 2/(T_i\tau_i) - \epsilon_i(\epsilon_i + 2\beta_i)/T_i^2$. Therefore, variations in the consumption rate are magnified under the instability condition

$$T_i > \epsilon_i\tau_i\,(\beta_i + \epsilon_i/2)\,. \tag{31}$$

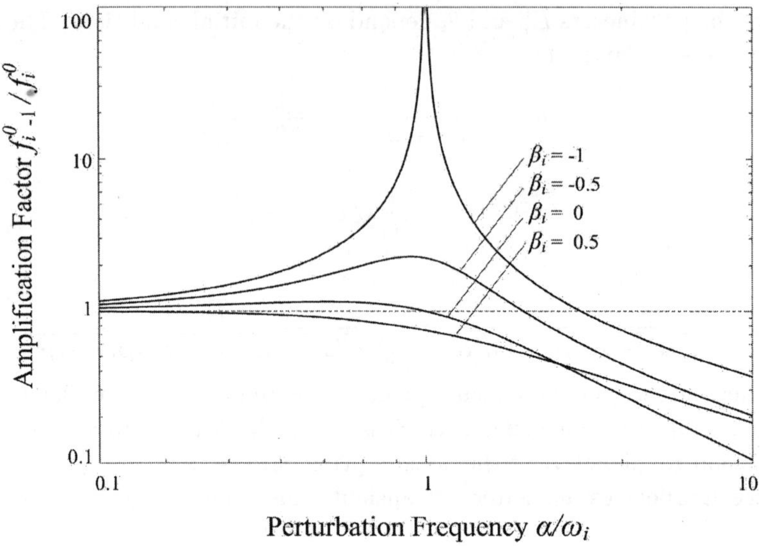

Fig. 2.2. Frequency response for different β_i and $T_i = \tau_i = \epsilon_i = 1$. For small β_i, corresponding to a small prognosis time horizon Δt, a resonance effect with an amplification factor greater than 1 can be observed. Perturbations with a frequency α close to the eigenfrequencies $\omega_i = 1/\sqrt{T_i\tau_i}$ are amplified and cause variations in stock levels and deliveries to grow along the supply chain. This is responsible for the bullwhip effect.

Supply chains show the bullwhip effect (which corresponds to the phenomenon of convective, i.e. upstream amplification), if the adaptation time T_i is too large, if there is no adaptation to some equilibrium production speed (corresponding to $\epsilon_i = 0$), or if the production management reacts too strong to deviations of the actual stock level N_i from the desired one N_i^0 (corresponding to a small value of τ_i), see Fig. 2.3. The latter is very surprising, as it implies that the strategy

$$\frac{dQ_i}{dt} = \frac{1}{T_i\tau_i}[N_i^0 - N_i(t)]\,, \tag{32}$$

which tries to maintain a constant work in progress $N_i(t) = N_i^0$, would ultimately lead to an undesireable bullwhip effect, given that production units are adjusted individually, i.e. in a decentralized way. In contrast, the management strategy

$$\frac{dQ_i}{dt} = \frac{1}{T_i}\Big\{ - \beta_i \frac{dN_i}{dt} + \epsilon_i[Q_i^0 - Q_i(t)]\Big\} \tag{33}$$

would avoid this problem, but it would not maintain a constant work in progress.

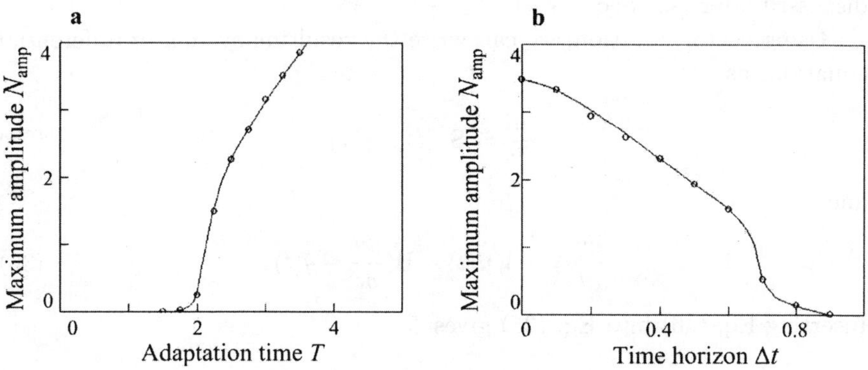

Fig. 2.3. (a) Plot of the maximum amplitude of oscillation in the inventories as a function of the adaptation time T. (b) Plot of the maximum amplitude of the inventories as a function of the time horizon Δt for an adaptation time of $T = 2$. (After Ref.[41])

The control strategy (32) with a sufficiently large value of τ_i would fulfill both requirements. Having the approximate relation (10) in mind, a forecast with a sufficiently long prognosis time horizon Δt (implying large values of β_i) is favourable for production stability. Delays in the determination of the inventories N_i, corresponding to negative values of Δt and β_i, are destabilizing. Note that the values of Δt which are sufficient to stabilize the system are often much smaller than the adaptation time T_i.[41]

2.3. Network Effects (with Péter Šeba)

The question is now, whether and how the bullwhip effect can be generalized to supply networks and how the dynamic behavior depends on the respective network structure. Linearization of the adaptation equation (6) leads to

$$\frac{dq_j}{dt} = -\sum_i V_{ji} n_i(t) - \sum_i W_{ji} \frac{dn_i}{dt} - \sum_l X_{jl} q_l(t) \,. \tag{34}$$

However, in order to avoid a vast number of parameters V_{ji}, W_{ji}, X_{jl} and to gain analytical results, we will assume product- and sector-independent parameters $V_{jk} = V\delta_{jk}$, $W_{jk} = W\delta_{jk}$, and $X_{jl} = \delta_{jl}$ in the following. This case corresponds to the situation that each production unit j is characterized by one dominating product $i = j$, which also dominates production control. For this reason, we additionally set $d_{ij} = \delta_{ij}$. Generalizations are discussed later (see Sec. 2.7.4).

Using vector notation, we can write the resulting system of differential equations as

$$\frac{d\vec{n}}{dt} = \mathbf{S}\vec{q}(t) - \vec{y}(t) \tag{35}$$

and

$$\frac{d\vec{q}}{dt} = -V\vec{n}(t) - W\frac{d\vec{n}}{dt} - \vec{q}(t)\,. \tag{36}$$

Inserting Eq. (36) into Eq. (35) gives

$$\frac{d}{dt}\begin{pmatrix}\vec{n}\\ \vec{q}\end{pmatrix} = \mathbf{M}\begin{pmatrix}\vec{n}\\ \vec{q}\end{pmatrix} + \begin{pmatrix}-\vec{y}\\ W\vec{y}\end{pmatrix} \tag{37}$$

with

$$\mathbf{M} = \begin{pmatrix}\mathbf{0} & , & \mathbf{S}\\ -V\mathbf{E} & , & -\mathbf{E} - W\mathbf{S}\end{pmatrix} \tag{38}$$

and the supply matrix $\mathbf{S} = \mathbf{D} - \mathbf{C} = (d_{ij} - c_{ij})$. For comparison of Eq. (19) the one above, one has to scale the time by introducing the unit time T/ϵ and to set $V = 1/(\tau\epsilon)$, $W = \beta/\epsilon$.

2.3.1. *General Methods of Solution*

It is possible to rewrite the system of $2u$ first order differential equations (37) in the form of a system of u differential equations of second order:

$$\frac{d^2\vec{q}}{dt^2} + (\mathbf{E} + W\mathbf{S})\frac{d\vec{q}}{dt} + V\mathbf{S}\vec{q}(t) = \vec{g}(t)\,, \tag{39}$$

where

$$\vec{g}(t) = V\vec{y}(t) + W\frac{d\vec{y}}{dt}\,. \tag{40}$$

Introducing the Fourier transforms

$$\vec{F}(\alpha) = \frac{1}{\sqrt{2\pi}}\int\limits_{-\infty}^{\infty} dt\; \mathrm{e}^{-\mathrm{i}\alpha t}\vec{q}(t) \tag{41}$$

and

$$\vec{G}(\alpha) = \frac{1}{\sqrt{2\pi}} \int_{-\infty}^{\infty} dt\, \mathrm{e}^{-\mathrm{i}\alpha t} \vec{g}(t) \tag{42}$$

reduces the problem to solving

$$\left[-\alpha^2 + \mathrm{i}\alpha(\mathbf{E} + W\mathbf{S}) + V\mathbf{S}\right] \vec{F}(\alpha) = \vec{G}(\alpha) \tag{43}$$

with given $\vec{G}(\alpha)$. The variable α represents the perturbation frequencies, and the general solution of Eq. (39) is given by

$$\vec{q}(t) = \frac{1}{\sqrt{2\pi}} \int_{-\infty}^{\infty} d\alpha\, \mathrm{e}^{\mathrm{i}\alpha t} \vec{F}(\alpha)\,. \tag{44}$$

Note that there exists a matrix $\mathbf{T}$ which allows one to transform the matrix $\mathbf{E} - \mathbf{S}$ via $\mathbf{T}^{-1}(\mathbf{E} - \mathbf{S})\mathbf{T} = \mathbf{J}$ into either a diagonal or a Jordan normal form $\mathbf{J}$. Defining $\vec{x}(\tau) = \mathbf{T}^{-1}\vec{q}(\tau)$ and $\vec{h}(t) = \mathbf{T}^{-1}\vec{g}(t)$, we obtain the coupled set of second-order differential equations

$$\frac{d^2 x_i}{dt^2} + 2\gamma_i \frac{dx_i}{dt} + {\omega_i}^2 x_i(t) = b_i \left[W x_{i+1}(t) + V \frac{dx_{i+1}}{dt} \right] + h_i(t)\,, \tag{45}$$

where

$$\gamma_i = [1 + W(1 - J_{ii})]/2\,, \quad \omega_i = [V(1 - J_{ii})]^{1/2} \quad \text{and} \quad b_i = J_{i,i+1}\,. \tag{46}$$

This can be interpreted as a set of equations for linearly coupled damped oscillators with damping constants γ_i, eigenfrequencies ω_i, and external forcing $h_i(t)$. The other forcing terms on the right-hand side are due to interactions of suppliers. They appear only if $\mathbf{J}$ is not of diagonal, but of Jordan normal form with some $J_{i,i+1} \neq 0$. Because of $b_i = J_{i,i+1}$, Eqs. (45) can again be analytically solved in a recursive way, starting with the highest index $i = u$. Note that, in the case $\mathbf{D} = \mathbf{E}$ (i.e. $d_{ij} = \delta_{ij}$), J_{ii} are the eigenvalues of the input matrix $\mathbf{C}$ and $0 \le |J_{ii}| \le 1$. Equation (45) has a special periodic solution of the form

$$x_i(t) = x_i^0\, \mathrm{e}^{\mathrm{i}(\alpha t - \chi_i)}\,, \tag{47}$$

$$h_i(t) = h_i^0\, \mathrm{e}^{\mathrm{i}\alpha t}\,, \tag{48}$$

where $\mathrm{i} = \sqrt{-1}$ denotes the imaginary unit. Inserting this into (45) and dividing by $\mathrm{e}^{\mathrm{i}\alpha t}$ immediately gives

$$(-\alpha^2 + 2\mathrm{i}\alpha\gamma_i + {\omega_i}^2) x_i^0 \mathrm{e}^{-\mathrm{i}\chi_i} = b_i(V + \mathrm{i}\alpha W) x_{i+1}^0 \mathrm{e}^{-\mathrm{i}\chi_{i+1}} + h_i^0\,. \tag{49}$$

With $\mathrm{e}^{\pm\mathrm{i}\phi} = \cos(\phi) \pm \mathrm{i}\sin(\phi)$ this implies

$$x_i^0 \mathrm{e}^{-\mathrm{i}\chi_i} = \frac{b_i(V + \mathrm{i}\alpha W) x_{i+1}^0 \mathrm{e}^{-\mathrm{i}\chi_{i+1}} + h_i^0}{-\alpha^2 + 2\mathrm{i}\alpha\gamma_i + {\omega_i}^2} = \frac{\sqrt{\mathrm{Re}^2 + \mathrm{Im}^2}\, \mathrm{e}^{\mathrm{i}\rho_i}}{\sqrt{({\omega_i}^2 - \alpha^2)^2 + (2\alpha\gamma_i)^2}\, \mathrm{e}^{\mathrm{i}\varphi_i}}\,, \tag{50}$$

where

$$\begin{aligned} \mathrm{Re} &= b_i x_{i+1}^0 [V\cos(\chi_{i+1}) + \alpha W \sin(\chi_{i+1})] + h_i^0\,, \\ \mathrm{Im} &= b_i x_{i+1}^0 [\alpha W \cos(\chi_{i+1}) - V\sin(\chi_{i+1})]\,, \end{aligned} \tag{51}$$

and

$$x_i^0 = \sqrt{\frac{[V^2 + (\alpha W)^2](b_i x_{i+1}^0)^2 + h_i^0 H_i + (h_i^0)^2}{({\omega_i}^2 - \alpha^2)^2 + (2\alpha\gamma_i)^2}} \tag{52}$$

with

$$H_i = 2 b_i x_{i+1}^0 [V\cos(\chi_{i+1}) + \alpha W \sin(\chi_{i+1})]\,. \tag{53}$$

Finally, we have

$$\chi_i = \varphi_i - \rho_i \quad \text{with} \quad \tan\varphi_i = 2\alpha\gamma_i/({\omega_i}^2 - \alpha^2) \tag{54}$$

and

$$\tan\rho_i = \frac{b_i x_{i+1}^0 [\alpha W \cos(\chi_{i+1}) - V \sin(\chi_{i+1})]}{b_i x_{i+1}^0 [V\cos(\chi_{i+1}) + \alpha W \sin(\chi_{i+1})] + h_i^0}\,. \tag{55}$$

For $h_i^0 = 0$, we obtain

$$\tan(\varphi_i - \chi_i) = \tan(\delta - \chi_{i+1}) \tag{56}$$

with

$$\tan\delta = \alpha W / V\,, \tag{57}$$

i.e. the phase shift between i and $i+1$ is just

$$\chi_i - \chi_{i+1} = \varphi_i - \delta\,. \tag{58}$$

According to Eq. (45), *the dynamics of our supply network model can be reduced to the dynamics of a sequential supply chain. However, the eigenvalues are now potentially complex and the new entities i have the meaning of "quasi-suppliers"* (analogously to "quasi-species"[46,47]) defined by the linear combination $\vec{x}(\tau) = \mathbf{T}^{-1}\vec{q}(\tau)$. This transformation makes it possible to define the bullwhip effect for arbitrary supply networks: It occurs if the amplitude x_i^0 is greater than the oscillation amplitude x_{i+1}^0 of the next downstream supplier $i+1$ and greater than the amplitude h_i^0 of the external forcing, i.e. if $x_i^0 / \max(x_{i+1}^0, h_i^0) > 1$.

Note that in contrast to sequential supply chains, the oscillation amplitude of $\vec{x}(t)$ may be amplified in the course of time, depending on the network structure. This case of *absolute instability* can occur if at least one of the eigenvalues $\lambda_{i,\pm}$ of the homogeneous equation (45) resulting for $h_i = b_i = 0$ has a positive real part, which may be true when some complex eigenvalues J_{ii} exist. The (up to) $2u$ eigenvalues

$$\begin{aligned}\lambda_{i,\pm} &= -\gamma_i \pm \sqrt{\gamma_i{}^2 - \omega_i{}^2} \\ &= -[1 + W(1 - J_{ii})]/2 \pm \sqrt{[1 + W(1 - J_{ii})]^2/4 - V(1 - J_{ii})} \qquad (59)\end{aligned}$$

depend on the (quasi-)supplier i and determine the temporal evolution of the amplitude of deviations from the stationary solution.

2.3.2. *Examples of Supply Networks*

It is useful to distinguish the following cases:

Symmetric supply networks: If the supply matrix $\mathbf{S}$ is symmetric, as for most mechanical or electrical oscillator networks, all eigenvalues J_{ii} are real. Consequently, if $\omega_i < \gamma_i$ (i.e. if V is small enough), the eigenvalues $\lambda_{i,\pm}$ of $\mathbf{M}$ are real and negative, corresponding to an overdamped behavior. However, if W is too small, the system behavior may be characterized by damped oscillations.

Irregular supply networks: Most natural and man-made supply networks have directed links, and $\mathbf{S}$ is not symmetric. Therefore, some of the eigenvalues J_{ii} will normally be complex, and an overdamped behavior is untypical. The characteristic behavior is rather of oscillatory nature (although asymmetry does not always imply complex eigenvalues[5], see Fig. 2.9). For small values of W, it can even happen that the real part of an eigenvalue $\lambda_{i,\pm}$ becomes positive. This implies an amplification of oscillations in time (until the oscillation amplitude is limited by non-linear terms). Surprisingly, this also applies to most upper triangular matrices, i.e. when no loops in the material flows exist.

Regular supply networks: Another relevant case are regular supply networks. These are mostly characterized by degenerate zero eigenvalues $J_{ii} = 0$ and Jordan normal forms $\mathbf{J}$, i.e. the existence of non-vanishing upper-diagonal elements $J_{i,i+1}$. Not only sequential supply chains, but also fully connected graphs, regular supply ladders, and regular distribution systems belong to this case[6] (see Fig. 2.4a, c, d). This is characterized by the two u-fold degenerate eigenvalues

$$\lambda_{\pm} = -(1+W)/2 \pm \sqrt{(1+W)^2/4 - V}\,, \tag{60}$$

independently of the suppliers i. For small enough values $V < (1+W)^2/4$, the corresponding supply systems show overdamped behavior, otherwise damped oscillations.

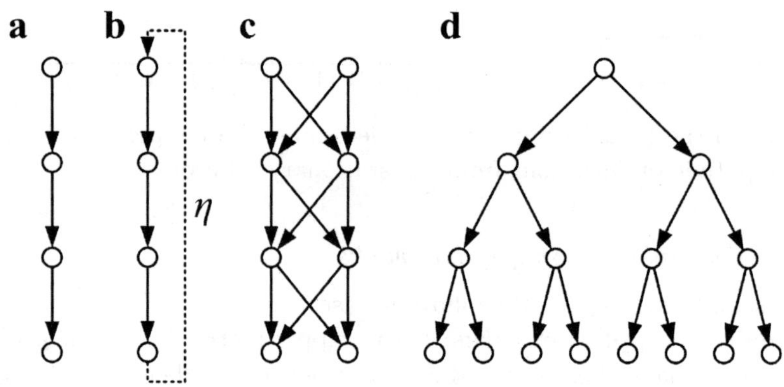

Fig. 2.4. Different examples of supply networks: (a) Sequential supply chain, (b) closed supply chain or supply circle, (c) regular supply ladder, (d) regular hierarchical distribution network.

For the purpose of illustration, the following equations display some regular input matrices **C** and their corresponding Jordan matrices **J**: For a fully connected network we have

$$\mathbf{C} = \begin{pmatrix} \frac{1}{u} & \frac{1}{u} & \frac{1}{u} & \frac{1}{u} & \frac{1}{u} & \ldots & \frac{1}{u} \\ \frac{1}{u} & \frac{1}{u} & \frac{1}{u} & \frac{1}{u} & \frac{1}{u} & \ldots & \frac{1}{u} \\ \frac{1}{u} & \frac{1}{u} & \frac{1}{u} & \frac{1}{u} & \frac{1}{u} & \ldots & \frac{1}{u} \\ \frac{1}{u} & \frac{1}{u} & \frac{1}{u} & \frac{1}{u} & \frac{1}{u} & \ldots & \frac{1}{u} \\ \frac{1}{u} & \frac{1}{u} & \frac{1}{u} & \frac{1}{u} & \frac{1}{u} & \ldots & \frac{1}{u} \\ \vdots & \vdots & \vdots & \vdots & \vdots & \ddots & \vdots \\ \frac{1}{u} & \frac{1}{u} & \frac{1}{u} & \frac{1}{u} & \frac{1}{u} & \ldots & \frac{1}{u} \end{pmatrix} \quad \text{and} \quad \mathbf{J} = \begin{pmatrix} 0 & 0 & 0 & 0 & 0 & \ldots & 0 \\ 0 & 1 & 0 & 0 & 0 & \ldots & 0 \\ 0 & 0 & 0 & 0 & 0 & \ldots & 0 \\ 0 & 0 & 0 & 0 & 0 & \ldots & 0 \\ 0 & 0 & 0 & 0 & 0 & \ldots & 0 \\ \vdots & \vdots & \vdots & \vdots & \vdots & \ddots & \vdots \\ 0 & 0 & 0 & 0 & 0 & \ldots & 0 \end{pmatrix}. \tag{61}$$

The Jordan normal matrix **J** of a sequential supply chain corresponds to the input matrix **C** itself, i.e. $\mathbf{J} = \mathbf{C}$. This, however, is quite exceptional.

For the supply ladder displayed in Fig. 2.4c we have

$$\mathbf{C} = \begin{pmatrix} 0 & 0 & \frac{1}{2} & \frac{1}{2} & 0 & 0 & 0 & 0 & 0 & 0 \\ 0 & 0 & \frac{1}{2} & \frac{1}{2} & 0 & 0 & 0 & 0 & 0 & 0 \\ 0 & 0 & 0 & 0 & \frac{1}{2} & \frac{1}{2} & 0 & 0 & 0 & 0 \\ 0 & 0 & 0 & 0 & \frac{1}{2} & \frac{1}{2} & 0 & 0 & 0 & 0 \\ 0 & 0 & 0 & 0 & 0 & 0 & \frac{1}{2} & \frac{1}{2} & 0 & 0 \\ 0 & 0 & 0 & 0 & 0 & 0 & \frac{1}{2} & \frac{1}{2} & 0 & 0 \\ 0 & 0 & 0 & 0 & 0 & 0 & 0 & 0 & \frac{1}{2} & \frac{1}{2} \\ 0 & 0 & 0 & 0 & 0 & 0 & 0 & 0 & \frac{1}{2} & \frac{1}{2} \\ 0 & 0 & 0 & 0 & 0 & 0 & 0 & 0 & 0 & 0 \\ 0 & 0 & 0 & 0 & 0 & 0 & 0 & 0 & 0 & 0 \end{pmatrix} \quad \text{and} \quad \mathbf{J} = \begin{pmatrix} 0 & 1 & 0 & 0 & 0 & 0 & 0 & 0 & 0 & 0 \\ 0 & 0 & 1 & 0 & 0 & 0 & 0 & 0 & 0 & 0 \\ 0 & 0 & 0 & 1 & 0 & 0 & 0 & 0 & 0 & 0 \\ 0 & 0 & 0 & 0 & 1 & 0 & 0 & 0 & 0 & 0 \\ 0 & 0 & 0 & 0 & 0 & 0 & 0 & 0 & 0 & 0 \\ 0 & 0 & 0 & 0 & 0 & 0 & 0 & 0 & 0 & 0 \\ 0 & 0 & 0 & 0 & 0 & 0 & 0 & 0 & 0 & 0 \\ 0 & 0 & 0 & 0 & 0 & 0 & 0 & 0 & 0 & 0 \\ 0 & 0 & 0 & 0 & 0 & 0 & 0 & 0 & 0 & 0 \\ 0 & 0 & 0 & 0 & 0 & 0 & 0 & 0 & 0 & 0 \end{pmatrix}, \tag{62}$$

where the number of ones corresponds to the number of levels of the supply ladder. For the hierarchical distribution network shown in Fig. 2.4d, but with 3 levels only, we have

$$\mathbf{C} = \begin{pmatrix} 0 & \frac{1}{2} & \frac{1}{2} & 0 & 0 & 0 & 0 \\ 0 & 0 & 0 & \frac{1}{2} & \frac{1}{2} & 0 & 0 \\ 0 & 0 & 0 & 0 & 0 & \frac{1}{2} & \frac{1}{2} \\ 0 & 0 & 0 & 0 & 0 & 0 & 0 \\ 0 & 0 & 0 & 0 & 0 & 0 & 0 \\ 0 & 0 & 0 & 0 & 0 & 0 & 0 \\ 0 & 0 & 0 & 0 & 0 & 0 & 0 \end{pmatrix} \quad \text{and} \quad \mathbf{J} = \begin{pmatrix} 0 & 1 & 0 & 0 & 0 & 0 & 0 \\ 0 & 0 & 1 & 0 & 0 & 0 & 0 \\ 0 & 0 & 0 & 0 & 0 & 0 & 0 \\ 0 & 0 & 0 & 0 & 1 & 0 & 0 \\ 0 & 0 & 0 & 0 & 0 & 0 & 0 \\ 0 & 0 & 0 & 0 & 0 & 0 & 1 \\ 0 & 0 & 0 & 0 & 0 & 0 & 0 \end{pmatrix}. \tag{63}$$

Note that the Jordan normal forms of different input matrices $\mathbf{C}$ may be identical, but the transformation matrix $\mathbf{T}$ and the driving term $\vec{h}(t)$ would then be different.

Randomized regular supply networks belong to the class of irregular supply networks, but they can be viewed as slightly perturbed regular supply networks. For this reason, there exist approximate analytical results for their eigenvalues. Even very small perburbations of the regular matrices $\mathbf{S}$ discussed in the previous paragraph can change the eigenvalue spectrum qualitatively. Instead of the two multiply degenerate eigenvalues $\lambda_\pm$ of Eq. (59), we find a scattering of eigenvalues around these values. The question is: why?

In order to assess the behavior of randomized regular supply networks, we apply Geršgorin's theorem on the location of eigenvalues.[48] According to this, the n complex eigenvalues $\lambda_k \in \mathbb{C}$ of some $n \times n$-matrix $\mathbf{N}$ are located in the union of n disks:

$$\lambda_k \in \bigcup_i \left\{ z \in \mathbb{C} : |z - N_{ii}| \leq \sum_{j(\neq i)} |N_{ij}| \right\} \tag{64}$$

Furthermore, if a union of l of these n discs form a connected region that is disjoint from all the remaining $n - l$ discs, then tere are precisely l eigenvalues of $\mathbf{N}$ in this region.

Let us now apply this theorem to the perturbed matrix

$$\mathbf{C}_\eta = \mathbf{C} + \eta \mathbf{P} \, . \tag{65}$$

For small enough values of η, *the corresponding eigenvalues* J_{ii} *should be located within discs of radius*

$$R_i(\eta) = \eta \sum_j |P_{ij}^{(\eta)}| \tag{66}$$

around the (possibly degenerated) eigenvalues J_{ii} *of the original matrix* $\mathbf{C}$ (see Fig. 2.5). This radius grows monotonously, but not necessarily linearly in the parameter η with $0 < \eta \leq 1$, which allows to control the size of the perturbation. Moreover, $\mathbf{P}^{(\eta)} = \mathbf{R}_\eta^{-1} \mathbf{P} \mathbf{R}_\eta$, where $\mathbf{R}_\eta$ is the orthogonal matrix which transforms $\mathbf{C}_\eta$ to a diagonal matrix $\mathbf{D}^{(\eta)}$, i.e. $\mathbf{R}_\eta^{-1} \mathbf{C}_\eta \mathbf{R}_\eta = \mathbf{D}^{(\eta)}$. (This assumes a perturbed matrix $\mathbf{C}_\eta$ with no degenerate eigenvalues.) Similar discs as for the eigenvalues of $\mathbf{C}_\eta$ can be determined for the associated eigenvalues $\lambda_{i,\pm}^{(\eta)}$ of the perturbed $(2u \times 2u)$-matrix $\mathbf{M}_\eta$ belonging to the perturbed $u \times u$-matrix $\mathbf{C}_\eta$, see Eq. (38) and Fig. 2.5.

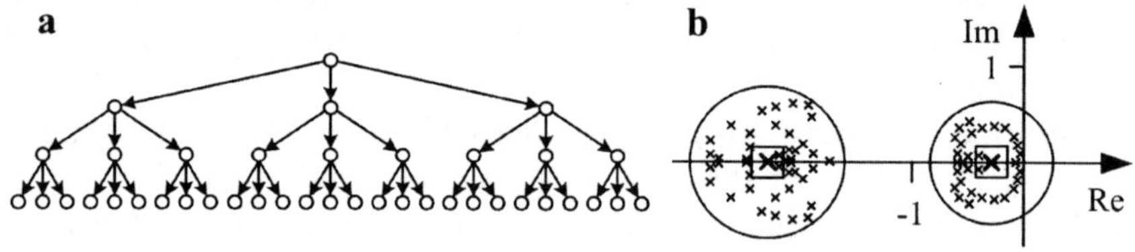

Fig. 2.5. (a) Example of a supply network (regular distribution network) with two degenerated real eigenvalues (see squares in the right subfigure). (b) The eigenvalues $\lambda_{i,\pm}$ of the randomly perturbed supply network (crosses) are mostly complex and located within Geršgorin's discs (large circles). (After Ref.[6])

Let us now discuss the example of a structural perturbation of a sequential supply chain (with $\eta = 0$) towards a supply circle (with $\eta = 1$), see Fig. 2.4b. For this, we set

$$\mathbf{C} = \begin{pmatrix} 0 & 1 & 0 & \ldots & 0 \\ 0 & 0 & 1 & \ldots & 0 \\ \vdots & & & \ddots & \vdots \\ 0 & 0 & 0 & \ldots & 1 \\ \eta & 0 & 0 & \ldots & 0 \end{pmatrix} . \tag{67}$$

While the normal form for $\eta = 0$ is given by a Jordan matrix $\mathbf{J}$ which agrees with $\mathbf{C}$, for any $\eta > 0$ we find the diagonal matrix

$$\mathbf{J} = \begin{pmatrix} J_{11} & 0 & \ldots & 0 \\ 0 & J_{22} & \ldots & 0 \\ \vdots & & \ddots & \vdots \\ 0 & 0 & 0 & J_{nn} \end{pmatrix} , \tag{68}$$

where the diagonal elements J_{ii} are complex and equally distributed on a circle of radius $\sqrt[n]{\eta}$ around the origin of the complex plane. Therefore, even an arbitrarily small perturbation can change the eigenvalues qualitatively and remove the degeneration of the eigenvalues.

2.4. Network-Induced Business Cycles (with Ulrich Witt and Thomas Brenner)

In order to investigate the macroeconomic dynamics of national economies, it is useful to identify the production units with economic sectors (see Fig. 2.6). It is also necessary to extend the previous supply network model by price-related equations. For example, increased prices $P_i(t)$ of products of sector i have a negative impact on the consumption rate $Y_i(t)$ and vice versa. We will describe this by a standard demand function L_i with a negative derivative $L_i'(P_i) = dL_i(P_i)/dP_i$:

$$Y_i(t) = [Y_i^0 + \xi_i(t)]L_i(P_i(t)) . \tag{69}$$

This formula takes into account random fluctuations $\xi_i(t)$ over time around a certain average consumption rate Y_i^0 and assumes that the average value of $L_i(P_i(t))$ is normalized to one. The fluctuation term $\xi_i(t)$ is introduced here in order to indicate that the variation of the consumption rate is a potentially relevant source of fluctuations.

Inserting (69) into (3) results in

$$\frac{dN_i(t)}{dt} = Q_i(t) - \sum_{j=1}^{m} c_{ij}Q_j(t) - \underbrace{[Y_i^0 + \xi_i(t)]L_i(P_i(t))}_{=Y_i(t)} . \tag{70}$$

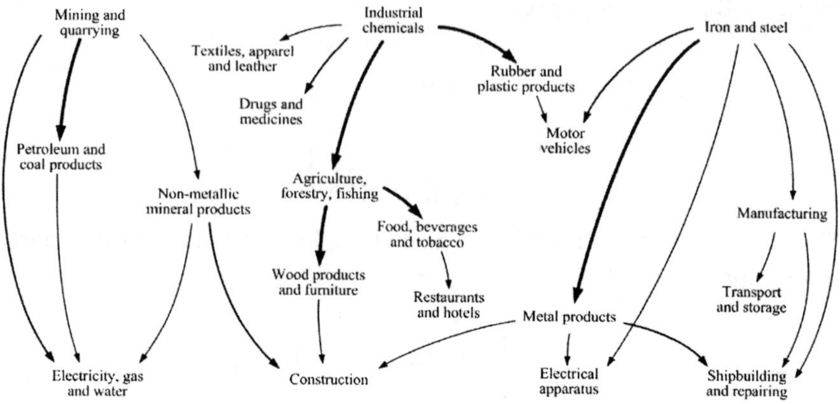

Fig. 2.6. Main service and commodity flows among different economic sectors according to averaged input-output data of France, Germany, Japan, UK, and USA. For clarity of the network structure, we have omitted the sector 'wholesale and retail trade', which is basically connected with all other sectors.

Herein, we have assumed $d_{ij} = \delta_{ij}$ as in Leontief's classical input-output model.[44] Moreover, in our simulations we have applied the common linear demand function

$$L_i(P_i) = \max(0, L_i^0 - L_i^1 P_i)\,, \tag{71}$$

where L_i^0 and L_i^1 are non-negative parameters.

Due to the price dependence of consumption, economic systems have the important equilibration mechanism of price adjustment. These can compensate for undesired inventory levels and inventory changes. A mathematical equation reflecting this is

$$\frac{1}{P_i(t)}\frac{dP_i}{dt} = \nu_i\left(\frac{N_i^0}{N_i(t)} - 1\right) - \frac{\mu_i}{N_i(t)}\frac{dN_i}{dt}\,. \tag{72}$$

The use of relative changes guarantees the required non-negativity of prices $P_i(t) \geq 0$. ν_i is an adaptation rate describing the sensitivity to relative deviations of the actual inventory $N_i(t)$ from the desired one N_i^0, and μ_i is a dimensionless parameter reflecting the responsiveness to relative deviations $(dN_i/dt)/N_i(t)$ from the stationary equilibrium state.

If the same criteria are applied to adjustments of the production rates $Q_i(t)$, we have the equation

$$\frac{1}{Q_i(t)}\frac{dQ_i}{dt} = \hat{\nu}_i\left(\frac{N_i^0}{N_i(t)} - 1\right) - \frac{\hat{\mu}_i}{N_i(t)}\frac{dN_i}{dt}\,. \tag{73}$$

$\hat{\alpha}_i = \hat{\nu}_i/\nu_i$ is the ratio between the adjustment rate of the output flow and the adjustment rate of the price in sector i. For simplicity, the same ratio $\hat{\alpha}_i = \hat{\mu}_i/\mu_i$ will be assumed for the responsiveness.

2.4.1. *Treating Producers Analogous to Consumers (with Dieter Armbruster)*

According to Eqs. (69) and (71), the change of the consumption rate in time is basically given by

$$\frac{dY_i}{dt} = [Y_i^0 + \xi_i(t)] \frac{dL_i}{dP_i} \frac{dP_i}{dt} = -[Y_i^0 + \xi_i(t)] L_i^1 \frac{dP_i}{dt} , \tag{74}$$

i.e. it is basically proportional to the price change dP_i/dt, but with a *negative* and potentially fluctuating prefactor. However,

$$\frac{1}{Q_i(t)} \frac{dQ_i}{dt} \propto \frac{1}{P_i(t)} \frac{dP_i}{dt} , \tag{75}$$

i.e. according to Eqs. (72) and (73), the change dQ_i/dt of the production rate close to the equilibrium state with $Q_i(t) \approx Q_i^0$ and $P_i(t) \approx P_i^0$ is basically proportional to the change dP_i/dt in the price with a *positive* prefactor. Is this an inconsistency of the model? Shouldn't we better treat producers analogous to consumers?

In order to do so, we have to introduce the delivery flows D_{ij} of products i to producers j. As in Eq. (74), we will assume that their change dD_{ij}/dt in time is proportional to the change dP_i/dt in the price, with a negative (and potentially fluctuating) prefactor. We have now to introduce the stock level $O_i(t)$ in the output buffer of producer i and the stock levels $I_{ij}(t)$ of product i in the input buffers of producer j. For the output buffer, we find the balance equation

$$\frac{dO_i}{dt} = Q_i(t) - \sum_j D_{ij}(t) - Y_i(t) \tag{76}$$

analogous to Eq. (70), as the buffer is filled with the production rate of producer i, but is emptied by the consumption rate and delivery flows. For the input buffers we have the balance equations

$$\frac{dI_{ij}}{dt} = D_{ij}(t) - c_{ij} Q_j(t) , \tag{77}$$

as these are filled by the delivery flows, but emptied with a rate proportional to the production rate of producer j, where c_{ij} are again the input coefficients specifying the relative quantities of required inputs i for production. Generalizations of these equations are discussed elsewhere.[38]

Let us now investigate the differential equation for the change dN_i/dt of the overall stock level of product i in all input and output buffers. We easily obtain

$$\frac{dN_i}{dt} = \frac{dO_i}{dt} + \sum_j \frac{dI_{ij}}{dt} = Q_i(t) - \sum_j c_{ij} Q_j(t) - Y_i(t) \,, \qquad (78)$$

as before. That is, the equations for the delivery flows drop out, and we stay with the previous set of equations for N_i, Q_i, P_i, and Y_i. As a consequence, we will focus on Eqs. (69) to (73) in the following.

2.5. Reproduction of Some Empirically Observed Features of Business Cycles

Our simulations of the above dis-aggregate (i.e. sector-wise) model of macroeconomic dynamics typically shows asynchronous oscillations, which seems to be characteristic for economic systems. Due to phase shifts between sectoral oscillations, the aggregate behavior displays slow variations of small amplitude (see Fig. 2.7). If the function $L_i(P_i)$ and the parameters $\nu_i/{\mu_i}^2$ are suitably specified, the non-linearities in Eqs. (70) to (69) will additionally limit the oscillation amplitudes.

Our business cycle theory differs from the dominating one[49,50] in several favourable aspects:

(i) Our theory explains irregular, i.e. non-periodic oscillations in a natural way (see Fig. 2.7). For example, w-shaped oscillations result as superposition of the asynchronous oscillations in the different economic sectors, while other theories have to explain this observation by assuming external perturbations (e.g. due to technological innovations).

(ii) Although our model may be extended by variables such as the labor market, interest rates, etc., we consider it as a potential advantage that we did not have to couple variables in our model which are qualitatively that different. Our model rather focusses on the material flows among different sectors.

(iii) Moreover, we will see that our model can explain emergent oscillations, which are not triggered by external shocks.

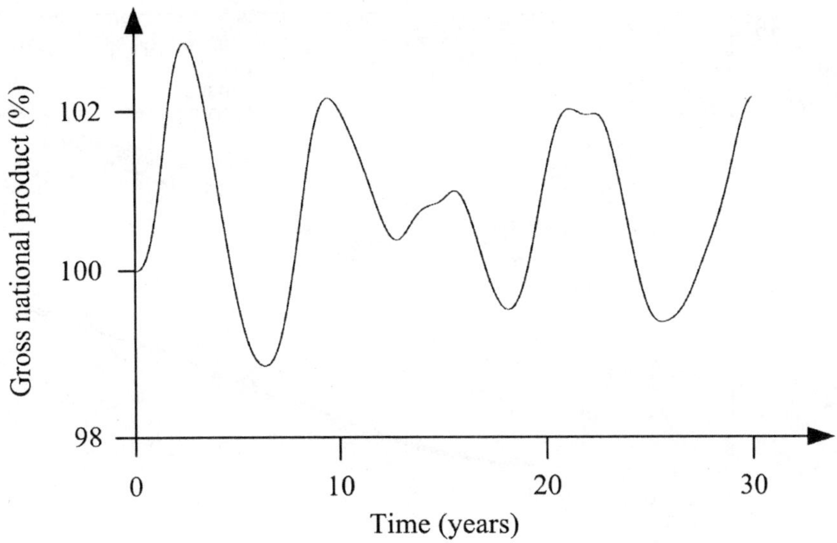

Fig. 2.7. Typical simulation result of the time-dependent gross domestic product $\sum_i Q_i(t)P_i(t)$ in percent, i.e. relative to the initial value. The input matrix was chosen as in Figs. 2.6 and 2.9a–d, but Y_i^0 was determined from averaged input-output data. Q_i^0 was obtained from the equilibrium condition, and the fluctuations $\xi_i(t)$ were specified as a Gaussian white noise with mean value 0 and variance $\sigma = 10.000$ (about 10% of the average final consumption). The initial prices $P_i(0)$ were selected from the interval [0.9;1.1]. Moreover, in this example we have assumed $L_i(P_i) = \max[0, 1 + d(P_i - P_i^0)]$ with $d = f'(P_i^0) = -L_i^1 = -10$ and the parameters $\nu_i = 0.1$, $\mu_i = 0.0001$, $\alpha_i = 1 = P_i^0$, and $N_i^0 = Y_i^0$. Although this implies a growth of small oscillations (cf. Fig. 1d), the oscillation amplitudes are rather limited. This is due to the non-linearity of model equations (70) to (69) and due to the phase shifts between oscillations of different economic sectors i. Note that irregular oscillations with frequencies between 4 to 6 years and amplitudes of about 2.5% are qualitatively well compatible with empirical business cycles. Our material flow model can explain w-shaped, non-periodic oscillations without having to assume technological shocks or externally induced perturbations. The long-term growth of national economies was intentionally not included in the model in order to separate this effect from network-induced instability effects. (After Ref.[5])

2.5.1. *Dynamic Behaviors and Stability Thresholds*

The possible dynamic behaviors of the resulting dis-aggregate macroeonomic model can be studied by analytical investigation of limiting cases and by means of a linear stability analysis around the equilibrium state (in which we have $N_i(t) = N_i^0$, $Y_i(t) = Y_i^0$, $Q_i^0 - \sum_j c_{ij} Q_j^0 = Y_i^0$, and $P_i(t) = P_i^0$). The linearized equations for the deviations $n_i(t) = N_i(t) - N_i^0$,

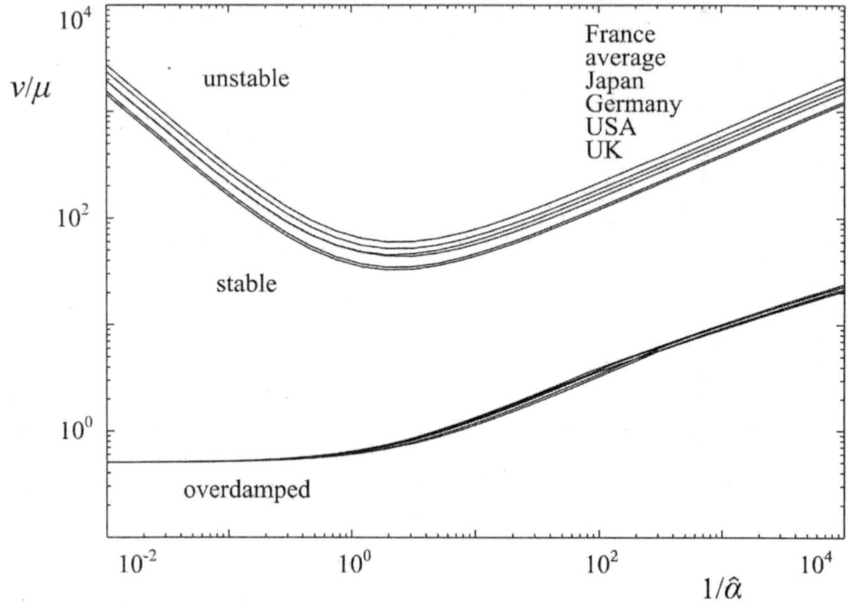

Fig. 2.8. The displayed phase diagram shows which dynamic behavior is expected by our dis-aggregate model of macroeconomic dynamics depending on the respective parameter combinations. The individual curves for different countries are a result of the different structures of their respective input matrices **C**. As a consequence, structural policies can influence the stability and dynamics of economic systems.

$p_i(t) = P_i(t) - P_i^0$, and $q_i(t) = Q_i(t) - Q_i^0$ from the equilibrium state read

$$\frac{dn_i}{dt} = q_i - \sum_j c_{ij} q_j - Y_i^0 f_i'(P_i^0) p_i - \xi_i(t)\,, \tag{79}$$

$$\frac{dp_i}{dt} = \frac{P_i^0}{N_i^0}\left(-\nu_i n_i - \mu_i \frac{dn_i}{dt}\right), \tag{80}$$

$$\frac{dq_i}{dt} = \frac{\hat{\alpha}_i Q_i^0}{N_i^0}\left(-\nu_i n_i - \mu_i \frac{dn_i}{dt}\right). \tag{81}$$

This system of coupled differential equations describes the response of the inventories, prices, and production rates to variations $\xi_i(t)$ in the demand. Denoting the m eigenvalues of the input matrix $\mathbf{C} = (c_{ij})$ by J_{ii} with $|J_{ii}| < 1$, the $3m$ eigenvalues of the linearized model equations are 0 (m

times) and

$$\lambda_{i,\pm} \approx \frac{1}{2}\Big(- A_i \pm \sqrt{(A_i)^2 - 4B_i}\Big) , \tag{82}$$

where

$$\begin{aligned} A_i &= \mu_i[C_i + \hat{\alpha}_i D_i(1 - J_{ii})] , \\ B_i &= \nu_i[C_i + \hat{\alpha}_i D_i(1 - J_{ii})] , \\ C_i &= P_i^0 Y_i^0 |f_i'(P_i^0)|/N_i^0 , \\ D_i &= Q_i^0/N_i^0 , \end{aligned} \tag{83}$$

see Fig. 2.8. Formula (82) becomes exact when the matrix $\mathbf{C}$ is diagonal or the parameters $\mu_i C_i$, $\hat{\alpha}_i \mu_i D_i$, $\nu_i C_i$ and $\hat{\alpha}_i \nu_i D_i$ are sector-independent constants, otherwise the eigenvalues must be numerically determined.

It turns out that the dynamic behavior mainly depends on the parameters $\hat{\alpha}_i$, $\nu_i/{\mu_i}^2$, and the eigenvalues J_{ii} of the input matrix $\mathbf{C}$ (see Fig. 2.9): In the case $\hat{\alpha}_i \to 0$ of fast price adjustment, the eigenvalues $\lambda_{i,\pm}$ are given by

$$2\lambda_{i,\pm} = -\mu_i C_i \pm \sqrt{(\mu_i C_i)^2 - 4\nu_i C_i} , \tag{84}$$

i.e. the network structure does not matter at all. We expect an exponential relaxation to the stationary equilibrium for $0 < \nu_i/{\mu_i}^2 < C_i/4$, otherwise damped oscillations. Therefore, immediate price adjustments or similar mechanisms are an efficient way to stabilize economic and other supply systems. However, any delay ($\hat{\alpha}_i > 0$) will cause damped or growing oscillations, if complex eigenvalues $J_{ii} = \mathrm{Re}(J_{ii}) + \mathrm{i}\,\mathrm{Im}(J_{ii})$ exist. *Note that this is the normal case, as typical supply networks in natural and man-made systems are characterised by complex eigenvalues* (see top of Fig. 2.9).

Damped oscillations can be shown to result if all values

$$\nu_i/{\mu_i}^2 = \hat{\alpha}_i \hat{\nu}_i/{\hat{\mu}_i}^2 \tag{85}$$

lie below the instability lines

$$\begin{aligned} \nu_i/{\mu_i}^2 \approx & \Big\{ C_i + \hat{\alpha}_i D_i [1 - \mathrm{Re}(J_{ii})] \Big\} \\ & \times \left(1 + \frac{\{C_i + \hat{\alpha}_i D_i[1 - \mathrm{Re}(J_{ii})]\}^2}{[\hat{\alpha}_i D_i \mathrm{Im}(J_{ii})]^2} \right) \end{aligned} \tag{86}$$

given by the condition $\mathrm{Re}(\lambda_{i,\pm}) \le 0$. For identical parameters $\nu_i/{\mu_i}^2 = \nu/\mu^2$ and $\hat{\alpha}_i = \hat{\alpha}$, the minimum of these lines agrees exactly with the numerically obtained curve in Fig. 2.9. Values above this line cause small oscillations to grow over time (see Appendix A).

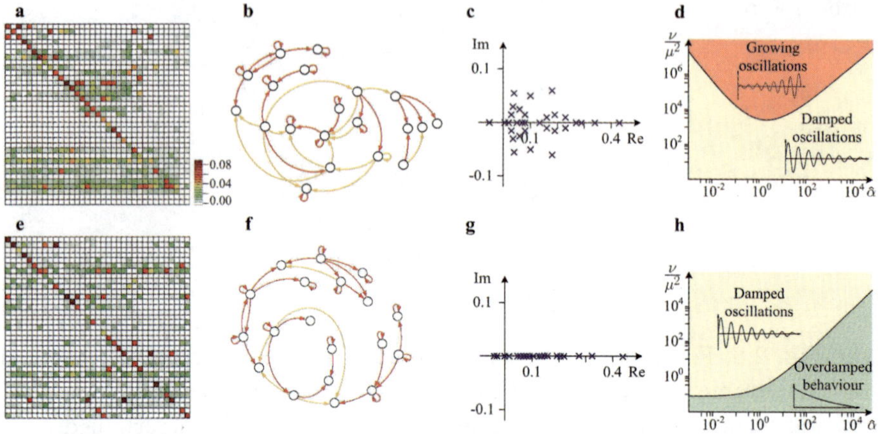

Fig. 2.9. Properties of our dynamic model of supply networks for a characteristic input matrix specified as average input matrix of macroeconomic commodity flows of several countries (top) and for a synthetic input matrix generated by random changes of input matrix entries until the number of complex eigenvalues was eventually reduced to zero (bottom). Subfigures (a), (e) illustrate the color-coded input matrices $\mathbf{A}$, (b), (f) the corresponding network structures, when only the strongest links (commodity flows) are shown, (c), (g) the eigenvalues $J_{ii} = \mathrm{Re}(J_{ii}) + \mathrm{i}\,\mathrm{Im}(J_{ii})$ of the respective input matrix $\mathbf{A}$, and (d), (h) the phase diagrams indicating the stability behavior of the model equations (70) to (69) on a double-logarithmic scale as a function of the model parameters $\hat{\alpha}_i = \hat{\alpha}$ and $\nu_i/{\mu_i}^2 = \nu/\mu^2 = V/M^2$. The other model parameters were set to $\nu_i = C_i = D_i = P_i^0 = N_i^0 = Y_i^0 = 1$. Surprisingly, for empirical input matrices $\mathbf{A}$, one never finds an overdamped, exponential relaxation to the stationary equilibrium state, but network-induced oscillations due to complex eigenvalues J_{ii}. (After Ref.[5])

In some cases, all eigenvalues J_{ii} of the input matrix $\mathbf{C}$ are real. This again applies to symmetric matrices $\mathbf{C}$ and matrices equivalent to Jordan normal forms. Hence, the existence of loops in supply networks is no sufficient condition for complex eigenvalues J_{ii} (see also Fig. 1f). It is also no necessary condition. For cases with real eigenvalues only, Eq. (82) predicts a stable, overdamped behavior if all values $\nu_i/{\mu_i}^2 = \hat{\alpha}_i\hat{\nu}_i/{\hat{\mu}_i}^2$ lie below the lines

$$\nu_i/{\mu_i}^2 \approx [C_i + \hat{\alpha}_i D_i(1 - J_{ii})]/4 \tag{87}$$

defined by $\min_i({A_i}^2 - 4B_i) > 0$ (see Appendix B). For identical parameters $\nu_i/{\mu_i}^2 = \nu/\mu^2$ and $\hat{\alpha}_i = \hat{\alpha}$, the minimum of these lines corresponds exactly to the numerically determined curve in Fig. 2.9. Above it, one observes damped oscillations around the equilibrium state, but growing oscillations are not possible. In supply systems with a slow or non-existent price adjustment mechanism (i.e. for $\hat{\alpha}_i \gg 1$ or $C_i = 0$), Eq. (87) predicts an overdamped behavior for real eigenvalues J_{ii} and

$$\hat{\nu}_i/\hat{\mu}_i{}^2 < D_i(1 - J_{ii})/4 \tag{88}$$

for all i, while Eq. (86) implies the stability condition

$$\hat{\nu}_i/\hat{\mu}_i{}^2 < D_i[1 - \mathrm{Re}(J_{ii})]\{1 + [1 - \mathrm{Re}(J_{ii})]^2/\mathrm{Im}(J_{ii})^2\} \tag{89}$$

for all i, given that some eigenvalues J_{ii} are complex. Moreover, for the case of sector-independent constants $V = \hat{\alpha}_i \nu_i D_i$ and $W = \hat{\alpha}_i \mu_i D_i$, the eigenvalues $\lambda_{i,\pm}$ can be calculated as

$$\lambda_{i,\pm} = -W(1 - J_{ii})/2 \pm \sqrt{[W(1 - J_{ii})]^2/4 - V(1 - J_{ii})}\,. \tag{90}$$

Considering that Eq. (36) for the adjustment of production speeds is slightly more general than specification (81), this result is fully consistent with Eq. (59).

2.6. Summary

In this contribution, we have investigated a model of supply networks. It is based on conservation equations describing the storage and flow of inventories by a dynamic variant of Leontief's classical input-output model. A second set of equations reflects the delayed adaptation of the production rate to some inventory-dependent desired production rate. Increasing values of the adaptation time T tend to destabilize the system, while increasing values of the time horizon Δt over which inventories are forecasted have a stabilizing effect.

A linear stability analysis shows that supply networks are expected to show an overdamped or damped oscillatory behavior, when the supply matrix is symmetrical. Some regular supply networks such as sequential supply chains or regular supply hierarchies with identical parameters show stable system behavior for all values of the adaptation time T. Nevertheless, one can find the bullwhip effect under certain circumstances, i.e. the oscillations in the production rates are larger than the variations in the consumption rate (demand). The underlying mechanism is a resonance effect.

Regular supply networks are characterized by multiply degenerate eigenvalues. When the related supply matrices are slightly perturbed, the degeneration is broken up. Instead of two degenerate eigenvalues, one finds complex eigenvalues which approximately lie on a circle around them in the complex plane. The related heterogeneity in the model parameters can have a stabilizing effect,[38] when they manage to reduce the resonance effect. Moreover, a randomly perturbed regular supply network may become linearly unstable, if the perturbation is large. If the supply matrix is irregular, the supply network is generally expected to show either damped or growing oscillations. In any case, supply networks can be mapped onto a generalized sequential supply chain, which allows one to define the bullwhip effect for supply networks.

While previous studies have focussed on the synchronization of oscillators in different network topologies,[8,51] we have found that *many supply networks display damped oscillations, even when their units—and linear chains of these units—behave in an overdamped way. Furthermore, networks of damped oscillators tend to produce growing (and mostly asynchronous) oscillations.* Based on these findings, it is promsing to explain business cycles as a result of material flows among economic sectors, which adjust their production rates in a decentralized way. Such a model can be also extended by aspects like labor force, money flows, information flows, etc.

2.7. Future Research Directions

2.7.1. *Network Engineering*

Due to the sensitivity of supply systems to their network structure, network theory[7,8,9,10,11,12,13,52] can make useful contributions: On the basis of Eqs. (86) and (87) one can design stable, robust, and adaptive supply networks ("network engineering"). For this reason, our present studies explore the characteristic features of certain types of networks: random networks, hierarchical networks, small world networks, preferential attachment networks, and others. In this way, we want to identify structural measures which can increase the robustness and reliability of supply networks, also their failure and attack tolerance. These results should, for example, be relevant for the optimization of disaster management.[53]

2.7.2. *Cyclic Dynamics in Biological Systems*

Oscillations are probably not always a bad feature of supply systems. For example, in systems with competing goals (such as pedestrian counterflows at bottlenecks[54] or intersecting traffic streams), oscillatory solutions can be favourable. Oscillations are also found in ecological systems[55] and some metabolic processes,[56,57,58,59] which could be also treated by a generalized model of supply networks. Examples are oscillations of nutrient flows in slime molds,[60] the citrate cycle,[61] the glycolytic oscillations in yeast,[56] and the Calcium cycle.[62,63,64] Considering the millions of years of evolution, it is highly unlikely that these oscillations do not serve certain biological functions. Apart from metabolic flows, however, we should also mention protein networks[59] and metabolic expression networks[57,58] as interesting areas of application of (generalized) supply network models.

2.7.3. *Heterogeneity in Production Networks*

Another interesting subject of investigation would be heterogeneity. The relevance of this issue is known from traffic flows[65] and multi-agent systems in economics[66]. Preliminary results indicate that, depending on the network structure, heterogeneous parameters of production units can reduce the bullwhip effect in supply networks.[38] That is, identical production processes all over the world may destabilize economic systems (as monocultures do in agriculture). However, in heterogeneous systems a lot more frequencies are influencing a single production step, which may also have undesireable effects. These issues deserve closer investigation.

2.7.4. *Multi-Goal Control*

Let us go back to Eq. (34) for the control of supply or production rates. This equation is to be applied to cases where a production unit produces several products and the production rate is not only adapted to the inventory of one dominating product, but also to the inventories of some other products. This leads to the problem of multi-goal control. It is obvious that the different goals do not always have to be in harmony with each other: A low inventory of one product may call for an increase in the production rate, but the warehouse for another product, which is generated at the same time, may be already full. A good example is the production of chemicals. We conjecture that multi-goal control of production networks may imply additional sources of dynamical instabilities and that some of the problems may be solved by price adjustments, which can influence the consumption rates. Again, a closer investigation of these questions is needed.

2.7.5. *Non-Linear Dynamics and Scarcity of Resources*

Finally, we should note that coupled equations for damped harmonic oscillators approximate the dynamic behavior of supply networks only close to their stationary state. When the oscillation amplitudes become too large, non-linearities will start to dominate the dynamic behavior. For example, they may select wave modes or influence their phases in favour of a synchronized behavior,[38,41] etc. Deterministically chaotic behavior seems to be possible as well.[30,67,68,69,70] Numerical results for limited buffer sizes and transport capacities have, for example, been presented by Peters *et al.*[71]

Additional non-linearities come into play, if there is a scarcity of resources required to complete certain products. Let's assume that the rate of transfering products i to j is Z_{ji}. If N_i is the number of products that may be delivered, the maximum delivery rate is therefore limited by $Z_{ji}N_i(t)$. The maximum production rate is given by the minimum of the delivery rate of all components i, divided by the respective number c_{ij} of parts required to complete one unit of product i. In mathematical terms, we have to make the replacement

$$Q_j(t) \longrightarrow Q_j(t) \min_i \left(1, \frac{Z_{ji}N_i(t)}{c_{ij}}\right) \tag{91}$$

in Eq. (70) and corresponding replacements in all derived formulas.[38] Therefore, the generalized relationship for systems in which resources may run short is

$$\frac{dN_i}{dt} = \sum_{j=1}^{u} (d_{ij} - c_{ij}) Q_j(t) \min_k \left(1, \frac{Z_{jk}N_k(t)}{c_{jk}}\right) - Y_i(t) \,. \tag{92}$$

These coupled non-linear equations are expected to result in a complex dynamics, if the transportation rate Z_{jk} or inventory N_k are too small. Note that this non-linearity may be even relevant close to the stationary state. In such cases, results of linear stability analyses for systems with scarce resources would be potentially misleading.

In situations where scarce resources may be partially substituted by other available resources, one may use the replacement

$$Q_j(t) \longrightarrow \left[1 + \sum_i \left(\frac{Z_{ji}N_i(t)}{c_{ij}}\right)^{z_j}\right]^{1/z_j} \tag{93}$$

instead of (91). Such an approach is, for example, reasonable for disaster management.[53] z_j is a parameter allowing to fit the ease of substitution of resources. The minimum function results in the limit $z_j \to -\infty$.

Acknowledgements

S.L. appreciates a scholarship by the "Studienstiftung des Deutschen Volkes". Moreover, the authors would like to thank Thomas Seidel for valuable discussions as well as SCA Packaging, the German Research Foundation (DFG projects He 2789/5-1, 6-1), and the NEST program (EU project MMCOMNET) for partial financial support.

Appendix A.

2.8. Boundary between Damped and Growing Oscillations

Starting with Eq. (82), stability requires the real parts $\text{Re}(\lambda_i)$ of all eigenvalues λ_i to be non-positive. Therefore, the stability boundary is given by $\max_i \text{Re}(\lambda_i) = 0$. Writing

$$C_i + \hat{\alpha}_i D_i(1 - J_{ii}) = \hat{\theta}_i + \text{i}\hat{\beta}_i \tag{A.1}$$

with $C_i = P_i^0 Y_i^0 |f_i'(P_i^0)|/N_i^0$ and defining

$$\begin{aligned} \hat{\theta}_i &= C_i + \hat{\alpha}_i D_i[1 - \text{Re}(J_{ii})]\,, \\ \hat{\beta}_i &= \mp\hat{\alpha}_i D_i \text{Im}(J_{ii}) \text{ (complex conjugate eigenvalues)}, \\ \hat{\gamma}_i &= 4\nu_i/{\mu_i}^2\,, \end{aligned} \tag{A.2}$$

we find

$$2\lambda_i/\mu_i = -\hat{\theta}_i - \text{i}\hat{\beta}_i + \sqrt{R_i + \text{i}I_i} \tag{A.3}$$

with

$$R_i = {\hat{\theta}_i}^2 - {\hat{\beta}_i}^2 - \hat{\gamma}_i\hat{\theta}_i \quad \text{and} \quad I_i = 2\hat{\theta}_i\hat{\beta}_i - \hat{\gamma}_i\hat{\beta}_i\,. \tag{A.4}$$

The real part of (A.3) can be calculated via the relation

$$\text{Re}\Big(\sqrt{R_i \pm \text{i}I_i}\Big) = \sqrt{\frac{1}{2}\Big(\sqrt{{R_i}^2 + {I_i}^2} + R_i\Big)}\,. \tag{A.5}$$

The condition $\text{Re}(2\lambda_i/\mu_i) = 0$ is fulfilled by $\hat{\gamma}_i = 0$ and

$$\hat{\gamma}_i = 4\hat{\theta}_i(1 + {\hat{\theta}_i}^2/{\hat{\beta}_i}^2)\,, \tag{A.6}$$

i.e. the stable regime is given by

$$\frac{\hat{\gamma}_i}{4} = \frac{\nu_i}{{\mu_i}^2} = \frac{\hat{\alpha}_i\hat{\nu}_i}{{\hat{\mu}_i}^2} \le \hat{\theta}_i\left(1 + \frac{{\hat{\theta}_i}^2}{{\hat{\beta}_i}^2}\right) \tag{A.7}$$

for all i, corresponding to Eq. (86).[5]

2.9. Boundary between Damped Oscillations and Overdamped Behavior

For $\hat{\alpha}_i > 0$, the imaginary parts of all eigenvalues λ_i vanish if $\text{Im}(J_{ii}) = 0$ (i.e. $\hat{\beta}_i = 0$) and if $R_i \geq 0$. This requires

$$\frac{4\nu_i}{{\mu_i}^2} = \hat{\gamma}_i \leq \hat{\theta}_i - \frac{{\hat{\beta}_i}^2}{\hat{\theta}_i} = \hat{\theta}_i = C_i + \hat{\alpha}_i D_i(1 - J_{ii}) \tag{A.8}$$

for all i, corresponding to Eq. (87).[5]

References

1. S. Tayur, R. Ganeshan, and M. J. Magazine, *Quantitative Models for Supply Chain Management* (Kluwer Academic, Dordrecht, 1998).
2. P. H. Zipkin, *Foundations of Inventory Management* (McGraw-Hill, Boston, 2000).
3. S. Nahmias, *Production and operations analysis* (McGraw-Hill/Irwin, Boston, 2001).
4. W. J. Hopp and M. L. Spearman, *Factory Physics* (McGraw-Hill, Boston, 2000).
5. D. Helbing, S. Lämmer, T. Brenner, and U. Witt, *Physical Review E*, **70**, 056118 (2004).
6. D. Helbing, S. Lämmer, T. Seidel, P. Šeba, and T. Płatkowski, *Physical Review E*, **70**, 066116 (2004).
7. D. J. Watts, S. H. Strogatz, *Nature* **393**, 440 (1998).
8. S. H. Strogatz, *Nature* **410**, 268 (2001).
9. R. Albert, A.-L. Barabási, *Reviews of Modern Physics* **74**, 47 (2002).
10. S. Maslov, K. Sneppen, *Science* **296**, 910 (2002).
11. M. Rosvall and K. Sneppen, *Physical Review Letters* **91**, 178701 (2003).
12. S. Bornholdt, H. G. Schuster, *Handbook of Graphs and Networks* (Wiley, Weinheim, 2003).
13. S. Bornholdt and T. Rohlf, *Physical Review Letters* **84**, 6114 (2000).
14. S. N. Dorogovtsev and J. F. F. Mendes *Evolution of Networks* (Oxford University Press, Oxford, 2004).
15. C. Daganzo, *A Theory of Supply Chains* (Springer, New York, 2003).
16. G. F. Newell, *Applications of Queueing Theory* (Chapman and Hall, Cambridge, U.K., 1982).
17. J. W. Forrester, *Industrial Dynamics* (MIT Press, Cambridge, MA, 1961).
18. F. Chen, Z. Drezner, J. K. Ryan, and D. Simchi-Levi, *Management Science* **46**(3), 436–443 (2000).
19. M. Baganha and M. Cohen, *Operations Research* **46**(3), 72-83 (1998).
20. J. Kahn, *Econom. Rev.* **77**, 667–679 (1987).
21. H. Lee, P. Padmanabhan, and S. Whang, *Sloan Management Rev.* **38**, 93–102 (1997).
22. H. L. Lee, V. Padmanabhan, and S. Whang, *Management Science* **43**(4), 546 (1997).

23. R. Metters, in *Proc. 1996 MSOM Conf.*, pp. 264–269 (1996).
24. J. Dejonckheere, S. M. Disney, M. R. Lambrecht, and D. R. Towill, *International Journal of Production Economics* **78**, 133 (2002).
25. J. Dejonckheere, S. M. Disney, M. R. Lambrecht, and D. R. Towill, *European Journal of Operational Research* **147**, 567 (2003).
26. S. M. Disney and D. R. Towill, *International Journal of Production Research* **40**, 179 (2002).
27. K. Hoberg, U. W. Thonemann, and J. R. Bradley, in *Operations Research Proceedings 2003*, Ed. D. Ahr, R. Fahrion, M. Oswald, and G. Reinelt (Springer, Heidelberg, 2003), p. 63.
28. B. Fait, D. Arnold, and K. Furmans, in *Operations Research Proceedings 2003*, Ed. D. Ahr, R. Fahrion, M. Oswald, and G. Reinelt (Springer, Heidelberg, 2003), p. 55.
29. E. Mosekilde and E. R. Larsen, *System Dynamics Review* **4**(1/2), 131 (1988).
30. J. D. Sterman, *Business Dynamics* (McGraw-Hill, Boston, 2000).
31. D. Armbruster, in *Nonlinear Dynamics of Production Systems*, Ed. G. Radons and R. Neugebauer (Wiley, New York, 2004), p. 5.
32. D. Marthaler, D. Armbruster, and C. Ringhofer, in *Proc. of the Int. Conf. on Modeling and Analysis of Semiconductor Manufacturing*, Ed. G. Mackulak *et al.* (2002), p. 365.
33. B. Rem and D. Armbruster, *Chaos* **13**, 128 (2003).
34. I. Diaz-Rivera, D. Armbruster, and T. Taylor, *Mathematics and Operations Research* **25**, 708 (2000).
35. D. Helbing, in *Traffic and Granular Flow '03*, Ed. S. Hoogendoorn, P. V. L. Bovy, M. Schreckenberg, and D. E. Wolf (Springer, Berlin, 2004), see e-print cond-mat/0401469.
36. A. Ponzi, A. Yasutomi, and K. Kaneko, *Physica A* **324**, 372 (2003).
37. U. Witt and G.-Z. Sun, in *Jahrbücher f. Nationalökonomie u. Statistik* (Lucius & Lucius, Stuttgart, 2002), Vol. 222/3, p. 366.
38. D. Helbing, *New Journal of Physics* **5**, 90.1–90.28 (2003).
39. D. Helbing, in *Nonlinear Dynamics of Production Systems*, Ed. G. Radons and R. Neugebauer (Wiley, New York, 2004), p. 85.
40. G. Radons and R. Neugebauer, Eds., *Nonlinear Dynamics of Production Systems* (Wiley, New York, 2004).
41. T. Nagatani and D. Helbing, *Physica A* **335**, 644 (2004).
42. T. Nagatani, *Physica A* **334**, 243 (2004).
43. D. W. Jorgenson, *The Review of Economic Studies* **28**(2), 105–116 (1961).
44. W. W. Leontief, *Input-Output-Economics* (Oxford University, New York, 1966).
45. E. Lefeber, Nonlinear models for control of manufacturing systems in *Nonlinear Dynamics of Production Systems*, Ed. G. Radons and R. Neugebauer (Wiley, New York, 2004), p. 69.
46. M. Eigen and P. Schuster, *The Hypercycle* (Springer, Berlin, 1979).
47. J. Padgett, Santa Fe Institute Working Paper No. 03-02-010.
48. R. A. Horn and C. R. Johnson, *Matrix Analysis* (Cambridge University, Cambridge, 1985).

49. J. B. Long jr. and C. I. Plosser, *Journal of Political Economy* **91** (1), 39 (1983).
50. G. W. Stadler, *Journal of Economic Literature* **32** (4), 1750 (1994).
51. L. F. Lago-Fernández, R. Huerta, F. Corbacho, and J. A. Sigüenza, *Physical Review Letters* **84**, 2758 (2000).
52. I. Stewart, *Nature* **427**, 601 (2004).
53. D. Helbing and C. Kühnert, *Physica A* **328**, 584 (2003).
54. D. Helbing and P. Molnár, *Physical Review E* **51**, 4282 (1995).
55. O. N. Bjørnstad, S. M. Sait, N. C. Stenseth, D. J. Thompson, M. Begon, *Nature* **409**, 1001 (2001).
56. S. Dane, P. G. Sørensen, and F. Hynne, *Nature* **402**, 320 (1999);
57. M. B. Elowitz and S. Leibler, *Nature* **403**, 335 (2000).
58. E. Almaas, B. Kovács, T. Vicsek, Z. N. Oltvai, and A.-L. Barabási, *Nature* **427**, 839 (2004).
59. A. S. Mikhailov and B. Hess, *J. Biol. Phys.* **28**, 655 (2002).
60. S. Koya and T. Ueda, *ACH Models in Chemistry* **135** (3), 297 (1998).
61. B. Alberts, A. Johnson, J. Lewis, M. Raff, K. Roberts, and P. Walter, *Molecular Biology of the Cell* (Garland Science, New York, 2002).
62. S. Schuster, M. Marhl, and T. Höfer, *Eur. J. Biochem.* **269**, 1333 (2002).
63. U. Kummer, L. F. Olsen, C. J. Dixon, A. K. Green, E. Bomberg-Bauer, and G. Baier, *Biophysical Journal* **79**, 1188 (2000).
64. L. Meinhold and L. Schimansky-Geier, *Physical Review E* **66**, 050901(R) (2002).
65. D. Helbing, *Reviews of Modern Physics* **73**, 1067 (2001).
66. M. Levy, H. Levy, and S. Solomon, *Journal de Physique I France* **5**, 1087 (1995).
67. T. Beaumariage and K. Kempf, in *Proceedings of the 5th IEEE/SEMI Advanced Semiconductor Manufacturing Conference*, p. 169 (1994).
68. T. Ushio, H. Ueda, and K. Hirai, *Systems & Control Letters* **26**, 335 (1995).
69. J. A. Hołyst, T. Hagel, G. Haag, and W. Weidlich, *Journal of Evolutionary Economics* **6**, 31 (1996).
70. I. Katzorke and A. Pikovsky, *Discrete Dynamics in Nature and Society* **5**, 179 (2000).
71. K. Peters, J. Worbs, U. Parlitz, and H.-P. Wiendahl, in *Nonlinear Dynamics of Production Systems*, Ed. G. Radons and R. Neugebauer (Wiley, New York, 2004), p. 39.

CHAPTER 3

MANAGING SUPPLY-DEMAND NETWORKS IN SEMICONDUCTOR MANUFACTURING

Karl Kempf

Decision Technologies Department, Intel Corporation
5000 W. Chandler Boulevard, Chandler, Arizona 85226, USA
E-mail: karl.g.kempf@intel.com

After describing the general supply-demand management problem with examples from the semiconductor industry, attention is restricted to the core manufacturing problem. Using a mathematical approach to this nonlinear stochastic combinatorial financial optimization problem, an outer loop for addressing the planning parts of the problem and an inner loop to manage the execution aspects are proposed. The outer loop provides a material release plan generated by a linear programming formulation (LP) and inventory safety stock targets generated by a dynamic programming formulation (DP) to the inner loop to guide execution. Portions of the nonlinearity and stochasticity inherent in the problem are addressed by the outer loop that requires iterative convergence between the LP and the DP. The inner loop is formulated from the perspective of model predictive control (MPC) and integrates concepts from optimal control and stochastic control. Results are presented to demonstrate the ability of the inner loop to track material release and safety stock targets while improving delivery performance in the face of both supply and demand stochasticity. A simulation module is also described that supports the other components of the system by validating their efficacy before application in the real world.

3.1. Introduction to Supply-Demand Networks

The economic systems that are the focus here stretch from the suppliers' suppliers through multiple manufacturing facilities to the customers' customers. Any high volume manufacturing company in the midst of this system plays many roles. On one hand, the company is a customer sending demand signals to upstream suppliers. To manufacture, the company needs raw materials relative to its products, production facilities and equipment including spares, skilled personnel, and often relies on subcontractors for burst capacity. In most cases, the subcontractor relies on the same materials and equipment suppliers as the company. Furthermore, the company's competitors often rely on this same set of suppliers.

On the other hand, the company is a supplier satisfying demand signals from downstream customers. In this role, the manufacturing company warehouses and transports (perhaps relying on logistics subcontractors) a variety of products to a range of geographically disperse consumers. These include other manufactures and their subcontractors, distributors, and end users, noting that the end users may also purchase from the distributors and the distributors might also purchase from the other manufacturers. Of course, the company's competitors sell to many of the same manufacturers, distributors, and end users. The tradition of referring to this supply-demand network as simply a "supply chain" grossly understates the actual complexity.

There are a broad set of flows inherent in this supply-demand network. Materials flow from suppliers to customers increasing in value while becoming products. Revenues move in the opposite direction through the various echelons in the network. Information flows in many directions including forecasts of supply towards the customers and forecasts of demand rolling up towards the suppliers. This provides a rich set of research and development opportunities for those interested in decision and control in networks of interacting entities.

Beyond the scope of this paper are applications of option theory and auction theory to the multiple interactions between suppliers and customers in the network, as well as applications of forecasting theory to the multiple interdependent supply and demand parameters of interest. Demand management issues, although extremely important, will not be addressed here. Neither will the extensive set of issues related to network design. The focus here will be on integrating optimal decision making and controlling decision execution in the manufacturing core of a supply-demand network. This business problem is addressed from the inside out under the hypothesis that, if the existing manufacturing core is not efficiently planned and executed, the probability of realizing efficient operation of the other components of the network becomes vanishingly small. Another reason for this focus is the ever-growing desire for mass customization and instantaneous doorstep delivery. Consumers have come to expect higher performance at lower prices improving on a year to year basis. In competition between supply-demand networks to satisfy these desires and expectations, success or failure rests to a large degree on the agility and responsiveness of the manufacturing core.

3.2. Examples from the Semiconductor Manufacturing

Examples will be drawn from Intel Corporation as an international company that represents the manufacturing core of a supply-demand network for computing and communications products (among others). While greatly simplified for the purposes of this paper, the examples represent 10's of billions of dollars in annual sales to 10's of millions of end customers for 10's of thousands of diverse products. The most sophisticated current semiconductor logic products integrate roughly 250 million transistors on a silicon die the size of an average human thumb print, and have continued to increase in complexity in accordance with Moore's law for over 30 years. The factories required to manufacture products of such sophistication current cost roughly 3 billion dollars to construct and outfit, and have continued to become more expensive with every generation of decreasing transistor size.

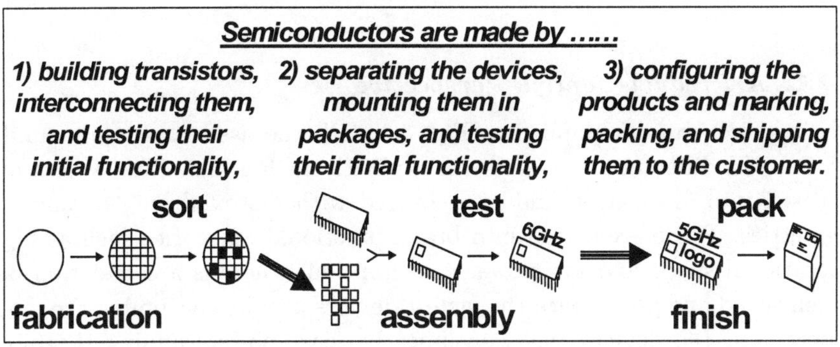

Fig. 3.1. The basic flow of semiconductor manufacturing.

The basic manufacturing flow is shown in Fig. 3.1. Transistors are fabricated on silicon wafers and interconnected to form circuits in a process that might consist of 300 individual production steps and take roughly 6 weeks to complete. The resulting wafers are tested to sort working die from those that do not function, and further sort the working die into broad functional categories. The factory output reflects a number of stochastic processes that underlie semiconductor manufacturing including random machine breakdowns that drive a distribution of throughput times (TPT) and random atomic misplacements that cause the resulting devices to exhibit a distribution of clock speeds and power consumptions.

Sorted wafers are then passed into the assembly process that might consist of 30 individual production steps and take a week or two to complete. Here the individual die are cut from the wafers and mounted in packages for protection and to make them suitable for incorporation in other products, often being mounted on printed circuit boards. Once packaged, they are stressed to induce infant mortality and tested for final classification into performance categories. Stochasticity again drives a TPT distribution as well as a distribution of end product characteristics.

Categorized products then enter the finish and pack process that involves roughly 10 processing steps that take only a few days to complete. One of the unfortunate asymmetries of semiconductor manufacturing is encountered here. Depending on the demand in the marketplace, fast devices can be configured to run more slowly, but slow devices can not be enticed to run faster. Once the final performance is configured, devices are individually labeled and packed in the appropriate medium for shipment and subsequent incorporation into other products.

3.2.1. *A Product-Centric Perspective*

The product fan-out implied in Fig. 3.1 with the associated opportunity for delayed differentiation is shown explicitly in Fig. 3.2. Raw wafers are released into fabrication and exit covered with die exhibiting a range of properties. These are sorted into broad functional categories such as high operational speed and low power consumption. There is a correlation between speed and power with the fastest devices usually consuming the most power, and the devices that consume the least power running the most slowly. This categorization is used in assembly to decide which die to put into what packages. In the case of microprocessors, die with the highest clock speed are placed into server packages while die with the lowest power consumption are placed into mobile packages. Both die types might be placed into desktop packages. Testing splits the performance distributions into finer categories, usually by maximum clock speed. In finish, these categories are used to fill demand for specific products. It is sometimes the case that the multiple splitting of multiple distributions results in different production flows giving the same end product (as is seen in the middle

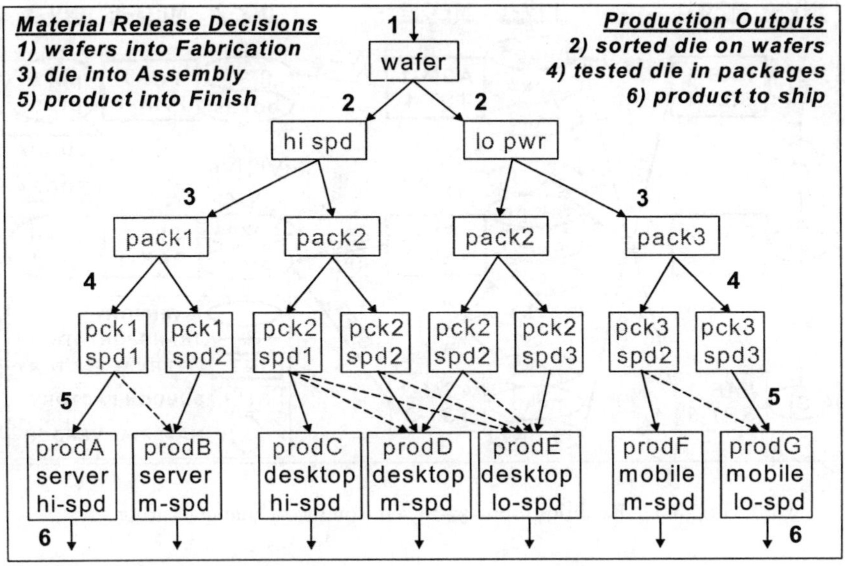

Fig. 3.2. Product fan-out in semiconductor manufacturing.

of Fig. 3.2). In addition, demand for lower speed devices can be filled by configuring higher speed devices with an associated lost opportunity cost.

3.2.2. *A Facilities-Centric Perspective*

The network of facilities alluded to in Fig. 3.1 with the associated opportunity for risk sharing is shown explicitly in Fig. 3.3. The first few echelons show vendors of silicon wafers and transportation links supplying raw material warehouses in front of fabrication (fab) facilities. Multiple fabrication/sort factories (FAB/SRT) and multiple vendors are involved to mitigate the risk of problems in any individual manufacturing facility. FAB/SRT facilities that are owned and operated by subcontractors are included as capacity buffers against variable demand. The middle few echelons show sorted die being transported to die warehouses in front of assembly (asm) facilities. In addition, vendors are supplying package warehouses through transportation links. Once again, multiple assembly/test factories (ASM/TST) and multiple vendors are involved as well as subcontractors. Notice the assignment of different die types and different package types to different factories. The last few echelons show tested product being transported to finished goods warehouses in front of finish (fin) facilities as well as

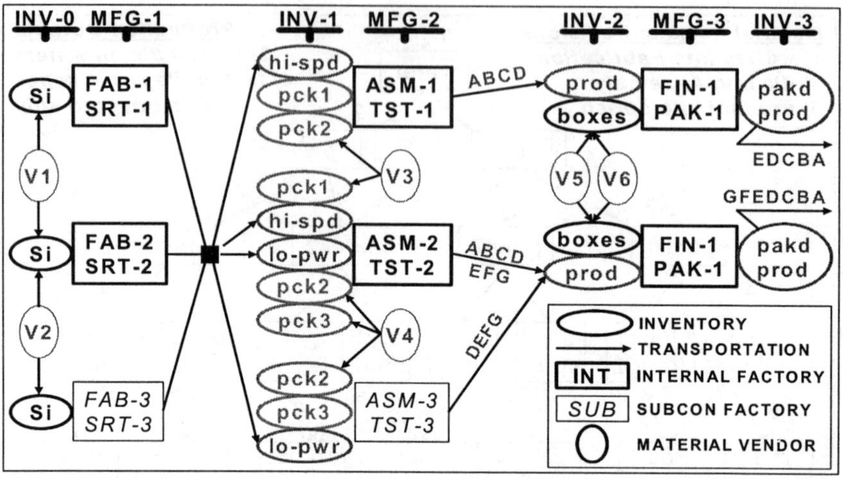

Fig. 3.3. Facilities topology in semiconductor manufacturing.

packing materials being supplied by vendors to pack facilities. In the example shown, multiple finish/pack factories (FIN/PAK) and multiple vendors are present but no subcontractors are involved. This is a reflection, in this example, of the primary manufacturer managing shipments to customers.

3.2.3. *Repetitive Decisions*

The core repetitive decisions that must be made during the operation of the supply-demand network are captured in these figures. Referring to Fig. 3.1, decisions must be made in every time period about how much of what material to release into fabrication, assembly, and finish facilities. In Fig. 3.2, how much material to put into which packages in assembly, and how much of what material to configure into which product in finish must be decided. Supporting these production decisions are a set of inventory decisions as shown in Fig. 3.3. One has to do with deciding how much of which raw material to hold in front of the fabrication (wafers), assembly (packages), and finish (boxes) facilities. Another deals with deciding how much of which work in progress to hold at assembly (die) and finish (unmarked product) factories.

Previous remarks focused attention in the larger set of supply-demand network problems to those dealing with the operation of the manufacturing core. This list of relevant decisions further refines the focus on issues discussed here. Excluded are the lower level details of internal factory decisions such as the setup and maintenance of machines and the batching and prioritization of lots that must be made in all of the factories represented in Fig. 3.3. Details of the operations of the warehouses and transportation links are similarly not considered. At a higher level, details of the decisions for managing the market place are excluded. This includes such important decisions as product introduction strategies and the timing of special sales offers. The decision system of interest here assumes a valid demand signal from above and a capable set of facilities below.

3.2.4. *Combinatorial Complexity*

There are a number of complications in the context of making these decisions. The most obvious from Figs. 3.2 and 3.3 is the combinatorial complexity of the problem. In actual practice, across the breadth of the offerings of an international high volume manufacturer, there could be as many as 25,000 distinct end products with the associated number of semi-finished goods, package types, and wafers. Across the globe, there could be as many as 100 factories and 500 inventory holding positions with the appropriate number of transportation links.

3.2.5. *Complexity from Supply Stochasticity*

A second complexity associated with Figs. 3.2 and 3.3 is supply stochasticity. Each of the factories manufacturing each of the products exhibits at least three types of variability based on different stochastic processes. The length of time it takes the raw material input to a factory to emerge as output, known as the throughput time (TPT), can be best described by a distribution. The underlying stochastic processes include the contention of lots flowing through the factory for production resources as well as the random unavailability of those resources. For example, machines experience breakdowns and operators take breaks. These TPT distributions are skewed since there are more events that can occur to slow a lot down and increase its TPT than there are events that can speed a lot up and decrease its TPT.

How much of the raw material released into the factory will emerge as output is also variable depending on a random set of unavoidable events that can occur. There are many examples of these stochastic processes. At a large scale compared with the transistors being fabricated, silicon wafers repeated heated and cooled in the production process occasionally experience thermal stress resulting in cracking with the accompanying loss of all die on the wafer since the wafer can not be further processed. At a small scale, contaminates from the manufacturing process can fall onto a wafer in such a way as to short circuit transistor interconnects causing a die to malfunction and be rejected at sort.

How the output from each factory will function is another example of supply variability. Clock speed and power consumption are among the main functional characterizations of semiconductors and both of these can best be described with distributions. The semiconductor manufacturing process can be thought of as the arrangement of atoms to form and interconnect transistors. Controlling exactly how many atoms are added or subtracted in the individual process steps and exactly how they are positioned relative to each other involve many stochastic processes. It is the detailed outcome of all of these processes for each die that generates the functional distributions.

3.2.6. *Complexity from Demand Stochasticity*

A third complexity is that of demand stochasticity, and it is challenging to try to describe the underlying random processes. Ultimately the end customer decides what products to purchase, and given the very large number of end customers and products, it is difficult if not impossible to forecast how much of which product will be purchased in which locations at what times because of the complex interplay of economic conditions, needs, and fashion. Compounding this situation is the competition in the marketplace between end product suppliers who differentiate based on form, function, price, and service (to mention but a few of the vectors). This translates into variability in the demand signal that is sent to the decision and control system of interest here. Orders are placed for semiconductor devices that include the specific product name and quantity as well as delivery time and place. Subsequent random events in the market result in requests to alter all of these parameters for existing orders (including in the extreme canceling the order), and the flow of materials through the supply-demand network must be altered to accommodate.

As can be seen in Figs. 3.2 and 3.3, the design of the network addresses some aspects of this demand variability. The fan-out of products in Fig. 3.2 supports delayed differentiation, final configuration being made only a few days before product is released into the logistics network. The multi-trajectory flow in Fig. 3.3 (including multiple vendors and subcontractors) supports risk sharing. In each echelon, multiple factories are making the same product and multiple products are being made in the same factory. Between echelons, the output of each factory is tied to the input of many downstream factories just as the input of each is tied to many upstream outputs. While all of these mappings can be changed over time with the dynamics of the business, the decision system under discussion here will have to manage all of the remaining stochasticity by appropriately utilizing the product fan-out and multi-trajectory flows.

3.2.7. *Complexity from Nonlinearity*

Magnifying the complexity of both the combinatorics and the supply-demand stochasticity is the fact that many of the key relationships are nonlinear. Manufacturing TPT, as well as TPT variability, increase nonlinearly as the utilization of manufacturing resources increase due to congestion. The probability of stock out decreases nonlinearity as the amount of safety stock inventory increases. The well-known price elasticity curve expresses the nonlinear relationship between demand and selling price. And the history of the semiconductor industry shows a nonlinear drop in selling prices of individual products over their life cycle (currently in the range of 6 months to 3 years). Notice as well that these nonlinearities interact. A drop in price could lead to an increase in demand. An increase in demand could lead to a higher safety stock target to protect against stock out. An increase in demand and inventory increases the load and congestion in the factory leading to an increase in TPT. An increase in manufacturing TPT could lead to a less responsive system leading to an increase in stock out probability and a decrease in demand. The decision system directing the supply-demand network must recognize and deal with these interacting nonlinearities.

3.2.8. *Financial Complexity*

The final complexity in the decision problem described here has to do with the financial aspects of the problem. The primary goal in operating a supply-demand network is to realize a profit, and this involves a number of conflicting objectives. The most obvious is the tug of war between minimizing cost and maximizing revenues. More subtle is the balance between maximizing profits now (perhaps risking future profits) and maximizing profits in the future (perhaps delaying or foregoing current profits). Making too little of a product or delivering it late as a result of striving to minimize current or future costs too often leads to decreased revenues through delayed payments, late penalties, lost sales, and (in the worst case) lost customers. Making too much of a product or introducing it before the market is ready as a result of trying to maximize revenue now and in the future too often leads to higher costs through holding charges and increased risk of obsolesce leading to write offs.

3.3. Managing Supply-Demand Networks

Given an international supply-demand network that operates around the clock, the resulting problem can be described as a continuous nonlinear stochastic combinatorial financial optimization. With the scale of international supply-demand networks, the difference between a near-optimal and a non-optimal solution can be worth hundreds of millions of dollars per year. In the case of Intel Corporation with roughly 35B$ in annual revenue, an improvement in supply-demand network operations that resulted in customers increasing their orders by only 3% would mean over 1B$ per annum in increased revenue.

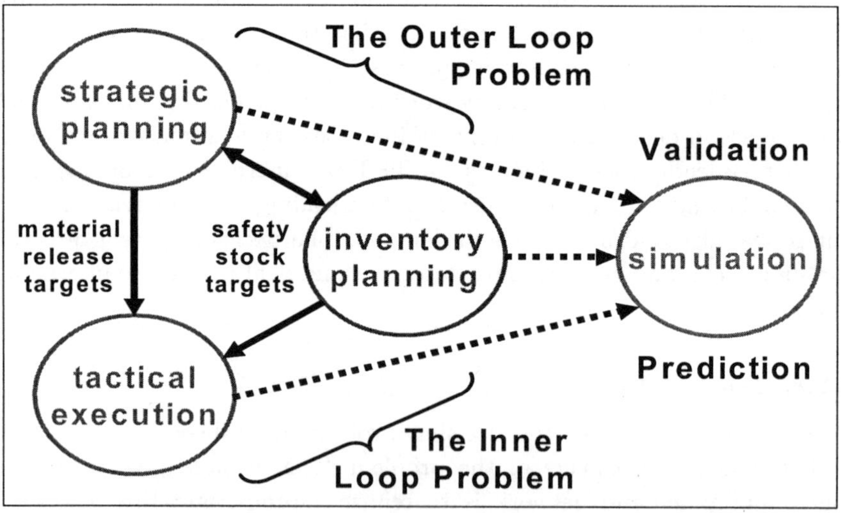

Fig. 3.4. Configuration of the proposed control system.

The approach described here relies on splitting the problem into a strategic planning function and a tactical execution function as shown in Fig. 3.4. The former can be thought of as an outer loop controller that considers business goals and trends over months and quarters into the future, and the latter as an inner loop controller that manages day to day operations providing the network with responsiveness and agility.

A major component of the strategic outer loop is a Linear Programming (LP) formulation that addresses the combinatorial financial optimization. The supply and demand nonlinearity and stochasticity is managed in two ways. On one hand, a major component of the tactical inner loop is a Model Predictive Control (MPC) formulation that accomplishes both feedback and feed forward control, selecting actions based on optimization of a control-relevant objective function. Stochasticity is contained by rapid measured responses as soon as deviation from the plan is recognized. On the other hand, a major component of both the outer and inner loops is an Inventory Planning formulation. The placement and sizing of safety stocks provides an additional hedge against both supply and demand stochasticity. From an outer loop perspective, safety stock targets are set based on long term demand forecasts that include potentially large errors and passed to the inner loop to guide execution. From an inner loop perspective, upper and lower control limits can be placed around the targets based on recent history in the execution environment and used to guide current execution. Simulation supports all of these components by providing for the testing of policies and plans before they are implemented in the actual supply-demand network. These components and interactions between them will be described in the next several sections.

3.3.1. *A Capacity Planning Formulation*

One solution to the strategic planning problem can be formulated as a mathematical optimization to allocate capacity to satisfy demand while minimizing costs and maximizing revenue.[1,2] There are three major categories of inputs in this approach as shown in Fig. 3.5.

One set of inputs specify the basic structure of the problem. This includes the material required to make any particular product as depicted in Fig. 3.2 and the possible manufacturing flows for products shown in Fig. 3.3. Supporting these descriptions are forecast future values of performance parameters including factory and transport TPTs and product yields. Financial data completes the set with the cost of manufacturing in each facility, transport costs for each link, the cost of holding inventory in each warehouse, and the average selling prices. Note that all of this data is time varying. New products are introduced and old products are

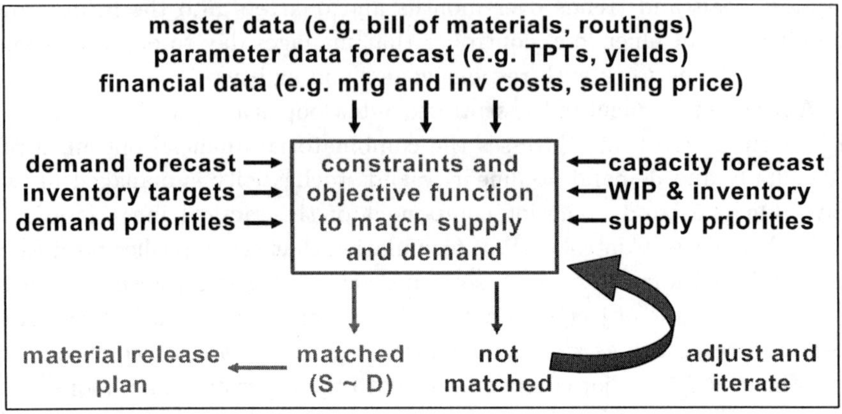

Fig. 3.5. A solution to the basic capacity planning problem.

discontinued with the associated modification of factory qualifications on a monthly basis. Yields and TPTs are intended to improve as products and factories mature over time. There is always external market pressure to lower selling prices and the associated internal pressure to reduce manufacturing costs. Note also that much of this data also varies by factory and product depending on the maturity of each. Different factories have different costs and performance parameters for manufacturing the same product. The same factory exhibits different performance parameters and costs for different products. Finally note that while TPTs and yields are the result of stochastic processes and are best described by distributions, only forecast means are used here.

The second set of inputs describes the supply scenario. Part of this description is the capacity forecast for all of the facilities in the system. This is product and facility specific and changes over time as individual facilities are modified. While there is a stochastic component to all of these capacity forecasts, again means are used. Another major part of the supply description is the current work in progress in each of the manufacturing facilities and transportation links as well as the current inventories in each warehouse. In practical applications there can be an arbitrary number of supply preferences included in the input. For example, it may be advantageous from a risk mitigation standpoint to distribute the load somewhat evenly across the supply facilities rather than heavily loading the low cost facility while leaving the higher cost facilities under-loaded. It is often challenging to include heuristic preferences in a mathematical formulation since they are usually difficult to quantify.

The demand scenario is described in the third set of inputs. The demand forecasts by product can vary dramatically over time and are a collection of numerical estimates on a highly stochastic process rooted in the dynamics of the marketplace. Included are safety stock targets intended to mitigate a portion of this demand stochasticity as well as part of the supply stochasticity. They are included on the demand side since they require capacity to be satisfied. The calculation of these targets will be the focus of a later section. Analogous to the supply scenario, an arbitrary number of demand preferences can be included. For example, a particular product or a particular customer may be deemed to have an elevated strategic importance relative to others although from a tactical financial perspective this may not be apparent. Once again, appropriate formulation is challenging but necessary to satisfy business needs.

Considering all of this information, the simplest formulation is based on mass balance and capacity constraints and an objective function that includes minimizing costs and maximizing revenues. The example formulation that follows includes only the left-most single flow leading to "prodA" from Fig. 3.2 and the top-most manufacturing facilities ("FAB-1 / SRT-1", "ASM-1 / TST-1", "FIN-1 / PAK-1") and product warehouses ("hi-spd", "prod", and "pakd prod") from Fig. 3.3. It ignores raw materials and transportation. Demand forecasts and inventory targets for the single product are included as are factory capacity, yield and TPT, initial factory WIP, and initial warehouse inventory for the single manufacturing line. The corresponding implied financial model includes manufacturing costs, inventory holding costs, penalties for missing safety stock targets and demand, and the average selling price. It ignores both demand and supply priorities.

Indices:

m	manufacturing stage (i.e. MFG-2) and sales
i	inventory location (i.e. INV-1)
p	product (intermediate or end product)
t	time

Input Variables (assumed non-negative):

$\text{Cap}_{m,p,t}$	total product that can be in a mfg stage
$\text{Yield}_{m,p,t}$	ave fraction of input to a mfg stage that exits

$\text{TPT}_{m,p,t}$	ave number of periods to complete a mfg stage
$\text{MfgCost}_{m,p,t}$	cost of manufacturing a product
$\text{InvTar}_{i,p,t}$	inventory target
$\text{InvCost}_{i,p,t}$	cost of holding inventory
$\text{InvPen}_{i,p,t}$	penalty for not satisfying inventory target
$\text{Dem}_{p,t}$	average demand
$\text{SaleBen}_{p,t}$	benefit from selling the end product
$\text{BackPen}_{p,t}$	penalty for not satisfying demand

Decision Variables (assumed non-negative):

$\text{Rel}_{m,p,t}$	material released into a mfg stage
$\text{Inv}_{i,p,t}$	inventory held at an inventory location
$\text{Sales}_{p,t}$	amount of satisfied demand
$\text{Backlog}_{p,t}$	amount of unsatisfied demand
$\text{InvUnd}_{i,p,t}$, $\text{InvOvr}_{i,p,t}$	slack variables, for deviation from inv target

Initialization:

$\text{Rel}_{m,p,t\in\{0-\text{TPT}_{\text{m,p,t}},0\}}$	previous material release
$\text{Inv}_{i,p,t=0}$	beginning on hand inventory
$\text{Backlog}_{p,t=0}$	no initial unsatisfied demand

Capacity Constraint

$$\sum_{t'\in(t-\text{TPT}_T,t]} \text{Rel}_{m,p,t'} \leq \text{Cap}_{m,p,t} \tag{1}$$

where $\text{T} = t - \text{TPT}_{m,p,T}$

Inventory Mass Balance Constraint

$$\text{Inv}_{i,p,t} = \text{Inv}_{i,p,t-1} + \sum_{T:T=t-\text{TPT}_{m,p,T}} (\text{Rel}_{m,p,T} \cdot \text{Yield}_{m,p,T}) - \text{Rel}_{m+1,p,t} \tag{2}$$

Backlog Mass Balance Constraint

$$\text{Backlog}_{p,t} = \text{Backlog}_{p,t-1} + \text{Dem}_{p,t} - \text{Sales}_{p,t} \tag{3}$$

where $\text{Sales}_{p,t} = \text{Rel}_{m=\text{Sales},p,t}$

Inventory Target Constraint

$$\text{Inv}_{i,p,t} + \text{InvUnd}_{i,p,t} - \text{InvOvr}_{i,p,t} = \text{InvTar}_{i,p,t} \tag{4}$$

Objective Function:

$$\max \left\{ \begin{aligned} &+\sum_{p,t} \text{Sales}_{p,t} \cdot \text{SaleBen}_{p,t} \\ &-\sum_{p,t} \text{Backlog}_{p,t} \cdot \text{BackPen}_{p,t} \\ &-\sum_{m,p,t} \text{Rel}_{m,p,t} \cdot \text{MfgCost}_{m,p,t} \\ &-\sum_{i,p,t} \text{Inv}_{i,p,t} \cdot \text{InvCost}_{i,p,t} \\ &-\sum_{i,p,t} (\text{InvUnd}_{i,p,t} + \text{InvOvr}_{i,p,t}) \cdot \text{InvPen}_{i,p,t} \end{aligned} \right\} \tag{5}$$

The primary result of executing such a formulation is a material release plan (or equivalently a product output plan) for all the manufacturing facilities included for each time segment in the horizon. Also available in the output is a transportation plan, an inventory plan including level relative to the specified targets, and a demand satisfaction plan including backlog. A profit projection is also available.

In an industrial setting, this tool is used for solidifying the response of the company to the market. It is rarely the case that initial executions provide a satisfactory match, usually leaving some demand unsatisfied and some capacity unused. In these circumstances, four of the inputs shown in Fig. 3.5 are manipulated. Selected demand can be moved earlier (incurring inventory holding costs), specific backlog can be authorized (incurring potential penalties), or some demand can be ignored (incurring lost revenue). Particular inventory targets can be adjusted (with attendant changes in inventory holding costs and penalties). Of course, demand and supply priorities can be modified potentially changing all facets of the plan. These manipulations reflect the practical impossibility of including all of the relevant business considerations in the formulation. Fortunately, a satisfactory strategic plan can usually be realized within a manageable number of iterations.

The principal shortcoming of this formulation is its treatment of the operational dynamics of the system. The most obvious is the disregard of the stochastic nature of the input parameter data, demand forecast, and capacity forecast. Addressing this deficiency is the focus of most of the rest of this paper. Another is the use of time periods with associated buckets of capacity to model the flow of work through manufacturing facilities. An example of the difficulties this precipitates has to do with the fact that TPT rises nonlinearly with utilization. To build the plan, fixed factory TPTs are input as parameters. The plan that results specifies the amount of material to be released into those factories, which determine their work loads and, in turn, the TPTs that will be realized.

One approach to this circularity has been to iteratively use an LP formulation and a discrete event model as described by Hung and Leachman3. Given an initial TPT, the LP provides an initial plan. This plan is executed in simulation to determine the resulting TPT. This TPT is fed back to the LP, and iteration is continued seeking convergence. Although convergence is not guaranteed, the results are generally adequate for most real-world applications. Unfortunately, solution time of such a scheme can be quite long if the simulation model contains the detail necessary to generate accurate TPTs.

More recently, a very efficient approach has been developed by Asmundsson, Rardin, and Uzsoy[4,5] using the idea of clearing functions. Here the expected throughput of a capacitated factory in a period of time is expressed as a function of its workload in that period. Based on an outer linearization of this nonlinear clearing function, an LP is formulated that appropriately captures the dynamics of factory capacity, TPT, and work load in a manner that supports rapid execution.

3.3.2. *An Inventory Planning Formulation*

One popular approach to managing the inherent stochasticity in the supply and in the demand is to put safety stock in place.[6,7,8] Such extra inventory hedges the risk of an unexpected supply downside or demand upside. The important decisions for each product include where to hold the safety stock as well as how much to hold, both varying over time. Consider the product and facilities topologies in Figs. 3.2 and 3.3, respectively. Positioning completed product in the final warehouse in the manufacturing flow so that it can be shipped from stock to the customer on demand tends to maximize revenues, but requires complete product differentiation and incurs inventory holding costs. Positioning undifferentiated material near the beginning of the manufacturing flow tends to minimize inventory holding costs, but requires customers to wait for the entire manufacturing TPT for their orders to be shipped. Holding too much extra inventory as safety stock increases holding costs and risks steep discounts or write-offs as the market becomes saturated or the products becomes obsolete. It can also waste manufacturing capacity, building what ultimately turns out to be the wrong product, potentially precluding building other more appropriate revenue products. Holding too little safety stock risks stock-outs with late delivery penalties, lost revenue, and in the worst case, lost customers. It can also stress the manufacturing system with rush orders and increased congestion leading to longer TPTs for all orders and the possibilities of lower yields.

Practical complications abound. For example, safety stock requirements change through the life cycle of products. In the ramp-up phase when market penetration is of paramount importance and stock-outs can not be tolerated, high levels of safety stock might be desirable but might be difficult to attain with immature production facilities. During the middle phase of market stability, two situations are possible based on the assumption that the initial market forecasts used to put capacity in place were flawed. If forecasts were pessimistic and the network is supply constrained, building safety stocks will be difficult since all capacity is being used to fill orders. If the network is demand constrained due to optimistic forecasts, the difficulty will be in overbuilding safety stock since manufacturing personnel loath idle capacity. In the ramp-down phase of a product's life cycle, safety stock is a liability when the focus is on moving customers to new improved products.

Furthermore, in operating supply-demand networks there is a substantial amount of product in motion and it can be difficult to identify that which is there for buffering variability. In an efficient network, the majority of the material flowing through manufacturing and transportation facilities is destined to satisfy firm customer orders. The majority of the material in warehouses should have a very low residency time as it moves steadily toward the market. However, some material might appear to be extra. Raw materials might have to be ordered in batches larger than can be immediately consumed. Intermediate production may arrive in a warehouse ahead of schedule due to manufacturing stochasticity in either higher yields or lower TPTs. Final product for actual customer orders may be built ahead due to mismatches between demand and supply at the future time that the order is due. None of these can be considered safety stock which must be scheduled into production facilities in addition to that committed to firm orders.

The simple supply-demand network described in Figs. 3.2 and 3.3 is among the most difficult for which to compute safety stock positions and amounts even for a bounded time horizon.[9,10] Multiple facilities exhibit a complex network flow with multiple products. Both supply and demand forecasts included substantial stochasticity with means that can vary over time. The flow contains multiple points at which multiple raw materials from outside vendors are injected. Fortunately the same product can be made through a number of routes, and the products incrementally differentiate along the flow providing opportunities for risk pooling.

One approach to the computation of safety stock positions and amounts is to construct a simulation that incorporates many of these complexities and then use it to search the space of possible positions and amounts for a good (but possibly not optimal) solution. Simulation speed and accuracy as well as the efficiency and effectiveness of the search control algorithm are crucial to the practicality of this approach. Glasserman and Tayur[11] have demonstrated a gradient estimation technique called infinitesimal perturbation analysis (IPA) for estimating from simulation the derivatives of inventory costs with respect to policy parameters. They have shown that they can use these derivatives to help steer the search for improved policies in multi-echelon systems with demand uncertainty related to those considered here.

Other approaches rely on representing the safety stock problem in such a way that mathematical optimization techniques can be employed. For example, Graves and Willems[12,13] have developed a method for supply-demand networks modeled as spanning trees that captures the stochastic nature of the demand and allows the safety stock problem, given a few simplifying assumptions, to be formulated as a mathematical optimization. The goal of this method is to place and size decoupling safety stocks that are large enough to permit downstream portions of a network to operate independently from the upstream, provided the upstream portion replenishes the external demand in a timely fashion. The simplifying assumptions include 1) bounded demand, 2) deterministic production TPTs at each stage that are independent of load (this is equivalent to assuming no capacity constraints), and 3) guaranteed service times by which each stage will satisfy its downstream demand. The first assumption is a practical one reflecting the fact that, to cover any possible demand eventuality however improbable, very large inventories would have to be positioned. Bounding demand simply means that, in extraordinary demand scenarios, the personnel operating the network would take extraordinary measures in response. The second assumption clarifies the focus on demand variability (but not supply variability) and inventory target setting (but not capacity allocation). The third assumption is key to the formulation. These service times for both end items and internal stages are decision variables in the optimization model used here including the possibility of setting a maximum service time for end items as required for customer satisfaction.

This safety stock planner models a supply-demand network as a set of nodes and arcs where nodes represent a processing function and arcs capture precedence relationship between nodes. While the LP formulation draws a distinction between manufacturing stages and inventory locations, here stages are defined as potential stocking locations and can hold safety stock after processing activity has been completed. The decision variables of a stage can be preset to prevent the holding safety stock. This safety stock optimization problem can be formulated as a mathematical program. The example formulation that follows includes only a single product flow from Fig. 3.2 and a single serial sequence through the production and warehouses facilities from Fig. 3.3. The financial model includes only inventory holding costs.

Indices:

i	stage

Input Variables (assumed non-negative):

$\mathrm{MaxDem}_i(t)$	max demand at stage i over an interval t
$\mathrm{AvgDem}_i(t)$	ave demand at stage i over an interval t
TPT_i	production lead time at stage i
$\mathrm{InvCost}_i$	cost of holding inventory at stage i
$\mathrm{MaxServ}_i$	max outgoing service time at stage i

Decision Variables (assumed non-negative):

$\mathrm{ServOut}_i$	outgoing service time from stage i
ServIn_i	incoming service time to stage i

Constraints:

$$\text{ServOut}_i - \text{ServIn}_i \le \text{TPT}_i \qquad \text{for all nodes} \tag{6}$$

$$\text{ServOut}_i \le \text{MaxServ}_i \qquad \text{for all nodes} \tag{7}$$

$$\text{ServIn}_i - \text{ServOut}_{i-1} \ge 0 \qquad \text{for all arcs} \tag{8}$$

Objective Function:

$$\min_{\text{ServOut}_i} \left\{ \begin{array}{l} \sum_i \text{InvCost}_i[\text{MaxDem}_i(\text{ServIn}_i + \text{TPT}_i - \text{ServOut}_i) \\ \quad -\text{AveDem}_i(\text{ServIn}_i + \text{TPT}_i - \text{ServOut}_i)] \end{array} \right\} \tag{9}$$

Generalizing for networks, the service times at a stage are the decision variables. Each stage quotes its downstream customers a guaranteed service time (ServOut_i) by which it will satisfy demand requests. The incoming service time to a stage (ServIn_i) is the maximum outgoing service time that its upstream supplier stages quote. The net replenishment time equals the incoming service time plus the production time (TPT) minus the outgoing service time. The net replenishment time τ at each stage dictates the inventory requirements at each stage to cover the demand over this time. The function $\text{MaxDem}_i(\tau)$ characterizes the maximum demand at each stage as a function of the net replenishment time. If a stage has a net replenishment time τ, the stage sets its base stock level equal to $\text{MaxDem}_i(\tau)$. The expected safety stock level at stage i will then equal the maximum demand minus the expected demand over the interval of length τ.

This formulation is solvable by Dynamic Programming (DP) where each stage solves a functional equation $\text{f}(\text{ServOut}_i)$ and/or $\text{g}(\text{ServIn}_i)$ depending on its orientation in the network. The DP seeks to determine the optimal set of service times that minimizes total safety stock cost while satisfying the maximum service time constraints to the final customer. The constraints on service time ensure that net replenishment times are nonnegative, incoming service time is not less than the maximum outgoing service time quoted to the stage, and the outgoing service times of demand stages do not exceeds the maximum service times imposed by customers.

3.3.3. *Integrating Capacity and Inventory Planning*

From a practical perspective, both the simulation and optimization approaches to inventory planning can raise philosophical questions. Their advocates suggest the computation of safety stock targets preceding the computation of the capacity plan on the grounds that the LP used for strategic planning expects inventory targets as input. The philosophical difficulty with this is rooted in the iterative business process use of the LP as described previously. That iteration is motivated by the desire to play out multiple supply and demand scenarios to find the one that best suits the overall strategy of the company. From this perspective, the role of safety stock is to protect the capacity plan that is ultimately selected from the relevant stochasticities that might disrupt it, and this can not be done a priori.

Note that both the stochastic simulation and dynamic programming approaches utilize some form of capacity allocation mapping between demand and production facilities during their computation. Notice also that the LP allocates specific demand to specific capacitated production facilities during its operation. This means that an iterative scheme might be appropriate. Two starting points are possible. In one, the LP is run first including heuristically set inventory targets (perhaps equal to current stocks or to zero), capacity allocations result that are passed to the safety stock computation for its first run, inventory targets result that are passed to the LP for a second run, and so on until convergence is attained. In the other, the iteration is initiated by first running the safety stock computation with heuristically set capacity allocations.

Initial investigations of these iterative schemes are in progress, and while promising, have identified an additional difficult decision problem. When the capacity of the supply system is greater than the sum of the demand forecast and the safety stock computed, convergence is relatively easily identified. In circumstances when either a) supply capacity is greater than demand forecast but less than demand forecast plus safety stock, or b) supply capacity is less than demand forecast and less than the sum of the demand forecast and the safety stock computed, convergence can not be resolved until additional capacity allocation choices are made. The decision is between building for demand or building for safety stock. This is especially interesting when deciding such comparisons as demand for lower margin products and safety stock for higher margin ones, or demand for products at the end of their life cycles and safety stock for ones early in theirs.

3.3.4. *A Tactical Execution Formulation*

The combination of capacity planning and inventory planning addresses a portion of the continuous nonlinear stochastic combinatorial financial optimization problem of concern here. But given the data preparation requirements as well as the business processes involved in using these tools, it is not likely that planning in a large international corporation will be practical or effective more often than once or twice per week. (Note that it is entirely possible that these tools will be used more often to explore supply and / or demand scenarios in the process of considering strategic options offline.) Unfortunately the nonlinear and stochastic aspects of a continuously operating supply-demand network are active minute to minute, hour by hour, day after day. It is clear that responding on a timescale much shorter than weekly will result in lower supply-demand network costs and improved levels of delivery performance generating higher revenues.

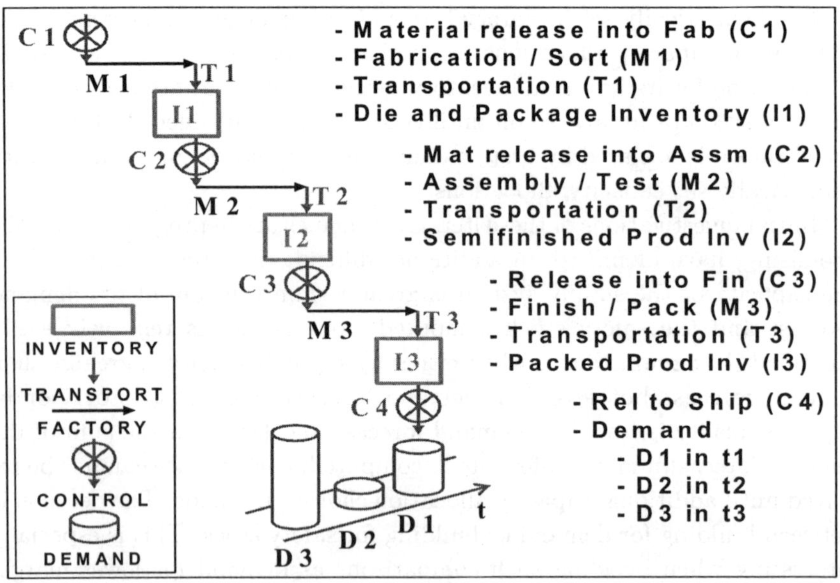

Fig. 3.6. A fluid analogy for semiconductor manufacturing.

The approach suggested is not more rapid execution of the outer loop tools, but rather relies on decision policies based on control-theoretic concepts applied to supply-demand networks. For more than 50 years, control methodologies have been continuously improved and reduced to reliable practice in a variety of process industries.[14] Process control systems are widely used to adjust flows to maintain product levels and compositions at desired levels. This is analogous to the management goals of high volume supply-demand networks and material flows in these networks can be modeled using a fluid analogy as shown in Fig. 3.6. In a very general sense, the manufacturing stages are represented as long and leaky "pipe" (to include TPT and yield, respectively) with the material in the pipes correspond to production work in progress. Additional pipes represent transportation links containing work in transit. Warehouses are represented as "holding tanks" and their contents correspond to inventory. Decisions about releasing material to initiate a production process or satisfy demand are implemented by adjusting control valves. Compare the system shown in Fig. 3.6 with the top-most facilities in Fig. 3.3. As a result of using this fluid analogy, one can expect that decision policies based on process control concepts will have a large and beneficial impact on supply-demand network management.

In particular, Model Predictive Control (MPC) offers a combined feedback-feed forward decision framework that can be tuned to provide enhanced performance and robustness in the presence of significant supply and demand variability and forecasting error while enforcing constraints on inventory levels as well as production and transportation capacities. Its formulation integrates optimal control, stochastic control, multivariable control, and control of processes with dead time. MPC is arguably the most general method currently known of posing the process control problem in the time domain.[15] In addition, there are early indications that MPC is applicable to the supply-demand network problems of interest here.[16–21]

In MPC, a system model along with current and historical measurements of the process are used to predict the system behavior at future time instants. A control-relevant objective function is then optimized to calculate a sequence of future control moves that must satisfy system constraints. The first predicted control move is implemented and, at the next sampling time, the calculations are repeated using updated system states. This is referred to as a Moving or Receding Horizon strategy and is illustrated in Fig. 3.7.

Input variables consist of two types: manipulate variables $\boldsymbol{u}$ which can be adjusted by the controller to achieve desired operation and disturbance or exogenous variables $\boldsymbol{d}$. The starts rates for F/S (C1), A/T (C2), and F/P (C3) represent manipulated variables for the problem in Fig. 3.6, with suggested targets determined by the capacity planning module. Demand

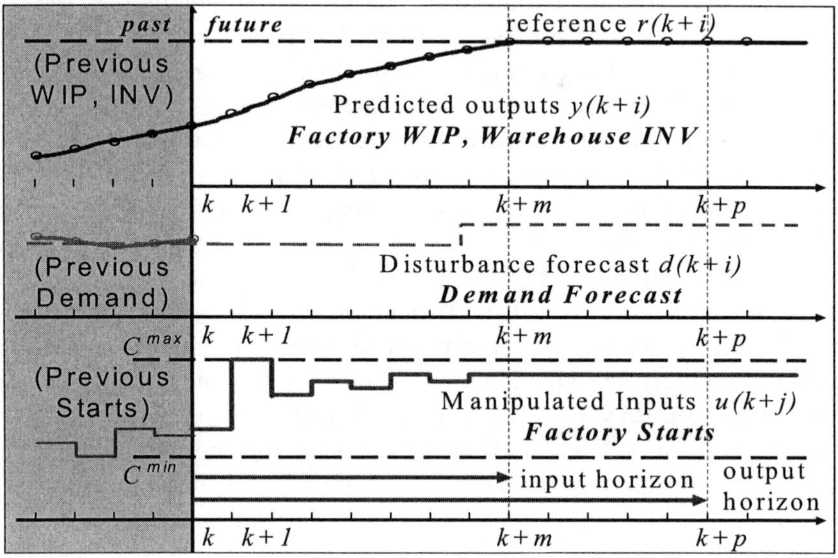

Fig. 3.7. The moving or receding horizon strategy.

(D1, D2, ...) in Fig. 3.6 is treated as an exogenous signal. This signal consists of actual demand which is only accurately known in the past and for a very short time into the future, and forecasted demand which is provided to all of the components shown in Fig. 3.4 by a separate organization such as Sales and Marketing. As noted in Fig. 3.7, the demand forecast is used in the moving horizon calculation to anticipate future system behavior and plays a significant role in the starts decisions made by the MPC controller. Representing quantities of primary importance to the system, $\boldsymbol{y}$ is a vector of output variables. Outputs can be classified in terms of controlled variables which must be maintained at some setpoint value and associated variables which may not have a fixed setpoint, but must reside between high and low limits. For the problem in Fig. 3.6, controlled variables consist of inventory levels (I1, I2, I3) whose setpoint targets are determined by the inventory planning module. Associated variables include loads on manufacturing nodes (M1, M2, M3) determined on the basis of their WIP.

In MPC control, predictions of y over a horizon are computed on the basis of an internal model arising from mass conservation relationships describing the dynamics of the manufacturing, inventory, and transportation nodes. For the problem in Fig. 3.6, the mass conservation relationship for inventories I can be written as:

$$I_x(k+1) = I_x(k) + C_x Y_{M_x}(k - \mathrm{TPT}_{M_x}) - C_{x+1}(k) \tag{10}$$

where x can equal 1, 2, or 3. For the manufacturing nodes an expression for the work in progress WIP is written as:

$$WIP_{M_x}(k+1) = WIP_{M_x}(k) + C_x(k) - C_x(k - \mathrm{TPT}_{M_x}) \tag{11}$$

where x can equal 1, 2, or 3, and TPT and Y represent the nominal throughput time and yield for the manufacturing node, respectively, while C represents the manufacturing starts per time period that constitute inflow for factories and outflow streams for warehouses. These systems of equations can, in general, be organized into a discrete-time state-space model representation amenable to Model Predictive Control implementation and analysis.[22]

The goal of the MPC decision policy is to seek a future profile for $\boldsymbol{u}$, the manipulated variables, that brings the system to some desired conditions consistent with the relevant constraints and the minimization of an objective function.

The ability to address constraints explicitly in the controller formulation is part of the appeal of MPC. For the problem in Fig. 3.6, constraints need to be imposed on the magnitudes of factory starts (12), the changes in factory starts (13), factory loads (14), and warehouse inventory levels (15).

$$0 \leq C_x(k+j|k) \leq C_x^{\max} \qquad x = 1 \text{ to } 3 \quad j = 1, 2, \ldots, m \tag{12}$$

$$\Delta C_x^{\min} \leq \Delta C_x(k+j|k) \leq C_x^{\max} \qquad x = 1 \text{ to } 3 \quad j = 1, 2, \ldots, m \tag{13}$$

$$0 \leq WIP_{M_x}(k+i|k) \leq CAP_{M_x}^{\max} \qquad x = 1 \text{ to } 3 \quad i = 1, 2, \ldots, p \tag{14}$$

$$I_x^{\min} \leq I_x(k+i|k) \leq I_x^{\max} \qquad x = 1 \text{ to } 3 \quad i = 1, 2, \ldots, p \tag{15}$$

While there is significant flexibility in the form of the objective function used in MPC, a meaningful formulation for the problem in Fig. 3.6 is:

$$\begin{aligned}\min_{\Delta C(k|k)\cdots\Delta C(k+m-1|k)} J = \sum_{x=1}^{3}\sum_{j=1}^{m} & Q_C(C_x(k+j-1|k) \\ & - C_{x:tar}(k+j-1|k))^2 \\ & + \sum_{x=1}^{3}\sum_{j=1}^{m} Q_{\Delta C}(\Delta C_x(k+j-1|k))^2 \\ & + \sum_{x=1}^{3}\sum_{i=1}^{p} Q_I(I_x(k+i|k) - I_{x:tar}(k+i))^2 \quad (16)\end{aligned}$$

The first input target term is meant to maintain the starts close to target values for each time period over the move horizon m based on the targets calculated by the outer loop capacity planner over time. The second move suppression term penalizes changes in the starts over the move horizon m. This term serves an important control-theoretic purpose as the primary means for achieving robustness in the controller in the face of uncertainty.[15] The third setpoint tracking term is intended to maintain inventory levels at targets specified by the outer loop inventory planner over time. These targets need not be constant and can change over the prediction horizon p. The emphasis given to each one of the sub-objectives is achieved through the choice of weights Q that can potentially vary over the move and prediction horizons.

For an MPC system relying on linear discrete-time state-space models to describe the dynamics, with an objective function as described above, and subject to linear inequality constraints, a numerical solution is achieved via a Quadratic Program (QP). Depending on the nature of the objective function, model and constraint equations, other programming approaches (LPs) may also be utilized.[23]

It is suggested that MPC-based formulations are able to perform satisfactorily if properly tuned in spite of the nonlinearities and stochasticity associated with semiconductor manufacturing supply-demand networks. This is illustrated for the representative problem described in Fig. 3.6. The major inputs to the controller for the experiment are shown in Table 3.1 including planned starts from the capacity planning system, inventory targets from the inventory planning system, and the demand forecast as used by both outer loop modules. The experimental model parameters for factories are shown in Table 3.2 including fixed factory capacity with stochastic TPTs and yields. Note that the TPTs are nonlinear with load for the fab/sort factory.

Table 3.1. Major Experimental Inputs.

	M_1	M_2	M_3
Planned Starts			
(units/day)	1,025	975	960
	*I*1	*I*2	*I*3
Inventory Targets			
(units)	2,700	1,550	1,450
	Ave		Var
Demand			
(units/day)	950		150
(uniform distribution)			

Demand can be measured and used to make predictions, but can not be manipulated. Planned starts can be manipulated by the controller and are modified to satisfy customer demand and keep inventories as close to target as possible while keeping factories within their allowed operational parameters.

The performance of the controller over a six month trial using $Q_C = 0$, $Q_{\Delta C} = 10$, $Q_I = 1$ is shown in Fig. 3.8. The factory starts in the left column have been adjusted relative to their targets in each time period while the loads in each factory in the middle column are appropriately managed. The load in the F/S factory is maintained at a high and stable level as desired for very expensive factories with long TPTs where "thrash" is likely to degrade performance. At the other end of the manufacturing flow, the F/P factory load varies from 30% of its maximum to 100% as expected in a low cost very short TPT factory that is tracking fluctuating demand. The A/T factory in the middle is absorbing supply stochasticity as well as demand stochasticity. The inventory levels in the right column are performing a similar function. In all three cases, the inventory averages are roughly at their target levels. But in each case there are times when levels range from near zero for a few days to brief periods at nearly double target levels. This is precisely what the safety stocks are intended to do, insulating manufacturing facilities from a large part of the variation in the

Table 3.2. Experimental Model Parameters.

			M_1	M_2	M_3
Capacity	(concurrent items	Max	45,000	7,500	2,500
	in factory)	Initial	33,500	5,700	1,900
TPT	(uniform distribution)				
		Min TPT (days)	30	5	1
	Load 0% to 70%	Ave TPT (days)	32	6	2
		Max TPT (days)	34	7	3
		Min TPT (days)	32	5	1
	Load 70% to 90%	Ave TPT (days)	35	6	2
		Max TPT (days)	38	7	3
		Min TPT (days)	35	5	1
	Load 90% to 100%	Ave TPT (days)	40	6	2
		Max TPT (days)	45	7	3
Yield	(uniform distribution)	Min Yield (%)	93	98	99
	(applied at factory exit)	Ave Yield (%)	95	99	99
		Max Yield (%)	97	99	100

system while assuring demand satisfaction. It is important to note that there was no backlog of customer orders during the six month experiment.

3.3.5. *Simulation Support*

Simulation can play many roles in the system described here for managing supply-demand networks. In each case, the speed with which the plan is generated, or the quality of the resulting plan, or the confidence in the plan producing the expected results (or some combination of these factors) is increased.

In the outer loop, there are multiple ways to utilize simulation. As described previously, using appropriate search control (such as Infinitesimal Perturbation Analysis[11]), a simulator can be run repeatedly to compute inventory targets. Or if inventory calculations are performed using mathematical optimization (for example, with Dynamic Programming[12,13]), a simulator that models the stochasticity of the real environment can be run

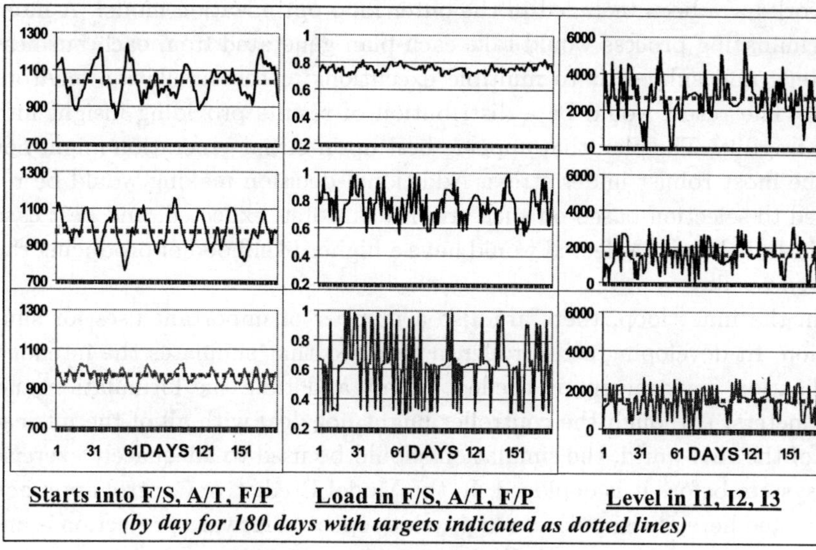

Fig. 3.8. The results of an MPC experiment.

to test whether the computed targets perform as expected. A similar test could be performed on the plan that results from running the capacity planning module alone or in tandem with the inventory planning module. This is similar in spirit to the iterative computation described previously between an LP model and a simulator to manage the nonlinearity between TPT and factory utilization.[3] The idea behind all of these activities is, if there is some question about whether or not the planning system has adequately modeled the nonlinearity and stochasticity in the execution environment, build a simulation of the execution environment and use it to evaluate the plan.

One interesting extension of this idea addresses the fragility of plans. Assume that a system has been implemented including a capacity planner and an inventory planner that iteratively generate a plan, and that an accurate simulation of the execution environment is available. Assume that the business problem being addressed is complex enough that a number of strategies have been investigated. It is tempting to believe that, since the solution to each strategy has been generated by mathematical optimization, the plan with the best objective function value is the plan that should be passed into the execution environment. Unfortunately this overlooks the fact that little if any of the dynamics of the supply-demand network have been included in the outer loop optimization model. A more discriminating process would take each plan generated from each business strategy and subject it to multiple executions in the simulation environment. The result would be a distribution of results providing insight into plan fragility. The plan that appears best upon initial generation might not be the most robust under stress. Additional decision making would be required to selection based on such criteria as worst, expected, and best case profit, but the plan selected would have a higher likelihood of producing the desired results.

In the inner loop, there are also a number of important uses for simulation. In developing a controller, a test bed that simulates the intended application environment is needed to test and tune the formulation and parameters. Although the controller might not deal with all of the intricacies of the real world, the simulation should be used to adequately exercise the system before it is deployed. In the Model Predictive Control approach advocated here,[15] a simulation to provide repeated forward projection is an inherent part of the method. Higher fidelity projection generates a number of benefits, although as shown in the experiment presented earlier, practically useful results are produced even from an approximate simulation.

This range of uses highlights one of the inherent tradeoffs that must be weighed in applying simulation to the supply-demand network problem. Generally, the higher the accuracy required of the simulation, the longer the computation time. On one hand, a very detailed discrete event simulation could be used internally in a MPC formulation to produce very high quality results for forward projection, but the run time of the resulting controller might compromise its usefulness. On the other hand, a very abstract fluid flow simulation could be used externally to the capacity and inventory planning modules to very quickly evaluate proposed plans, but the lack of detail might compromise the desired discriminating power.

Recent efforts have focused on improving this tradeoff. Historical applications of simulation to manufacturing have included fine details of the production process as well as the equipment and personnel to address problems of factory design and operation. Including this level of detail when considering the set of manufacturing, warehousing, and transportation entities involved in a supply-demand network would raise computation times well beyond practical limits. To address this difficulty, there have been explorations to find the simplest possible discrete event elements that can be coupled to provide a sufficiently accurate simulation for these networks.[24,25] Another approach to modeling is fluid networks as used, for example, in traffic theory.[26] Expanding this type of modeling to address the specific features important to simulating supply-demand networks has been explored recently.[26–30]

Whatever the simulation approach used, robustly interfacing the components shown in Fig. 3.4 is a challenge. The simulation component is present as a means to accurately model the material flow in the system. The other components are all involved in the decision flow required to manage the network. Algorithms for modeling material flow, whether based on differential equations or discrete events simulations, are very different from the optimization and control algorithms used to model decision flow. From one perspective, interfacing these algorithms is equivalent to modeling the data and information flow throughout the network. Recent work in this area has focused on a versatile methodology to interconnect a wide variety of simulation approaches and decision approaches based on the computing principles of model composability and system interoperability.[31,32]

3.4. Conclusions

Efficient operation of a complex supply-demand network can provide a company with a substantial competitive advantage. Unfortunately the planning problem for the manufacturing core of a network requires continuous nonlinear stochastic combinatorial financial optimization. One of the best ways to attack such a difficult problem is from a combined optimization and control perspective. An outer loop combining capacity planning and inventory planning can address the combinatorial and financial aspects of the problem. An inner loop based on Model Predictive Control can take material release and inventory target plans from the outer loops (with inventory control limits from recent execution results) and provide excellent customer service over long periods of time in the presence of nonlinearities and stochasticities. Multiple simulation techniques can provide support for each of these activities. Integrating these modules as a planning and execution system promises practical solutions to this very difficult but very important economic problem.

Acknowledgments

The author thanks John Bean of the Decision Technologies Group at Intel Corporation for the linear programming formulation, Sean Willems of the School of Management at Boston University for the dynamic programming formulation, and Wenlin Wang of the Department of Chemical and

Materials Engineering at Arizona State University for the quadratic programming formulation. The author also acknowledges Jakob Asmundsson, John Bean, Amit Devpura, Gary Godding, Michael O'Brian, Shamin Shirodkar, and Kirk Smith of Intel Corporation and Dieter Armbruster, Daniel Rivera, and Hessam Sarjoughian of Arizona State University for the stimulating interactions that have generated the approach to managing the core manufacturing component of a complex supply-demand network described here.

References

1. W. J. Hopp and M. L. Spearman, *Factory Physics: Foundations of Manufacturing Management*, New York: McGraw Hill, 1996, Ch. 16 (Aggregate and Workforce Planning).
2. S. Chopra and P. Meindl, *Supply Chain Management: Strategy, Planning, and Operation*, Upper Saddle River, NJ: Prentice-Hall, 2001, Part 2 (Planning Demand and Supply in a supply-demand network).
3. Y.-F. Hung and R. C. Leachman, "A production planning methodology for semiconductor manufacturing based on iterative simulation and linear programming calculations," *IEEE Trans. on Semiconductor Manufacturing*, Vol. 9, No. 2, pp. 257–269, 1996.
4. J. Asmundsson, R. L. Rardin, and R. Uzsoy, "Tractable nonlinear capacity models for production planning part I: modeling and formulation," *Operations Research*, submitted for publication.
5. J. Asmundsson, R. L. Rardin, and R. Uzsoy, "Tractable nonlinear capacity models for production planning part II: implementation and computational experiments," *Operations Research*, submitted for publication.
6. H. L. Lee and C. Billington, "Managing supply chain inventories: pitfalls and opportunities," *Sloan Management Review*, Vol. 33, pp. 65–73, Spring 1992.
7. W. J. Hopp and M. L. Spearman, *Factory Physics: Foundations of Manufacturing Management*, New York: McGraw Hill, 1996, Ch. 17 (Inventory Management).
8. S. Chopra and P. Meindl, *Supply Chain Management: Strategy, Planning, and Operation*, Upper Saddle River, NJ: Prentice-Hall, 2001, Part 3 (Planning and Managing Inventories in a supply-demand network).
9. S. C. Graves,"Safety stocks in manufacturing systems," J. Manufacturing and Operations Management, Vol. 1, pp. 67–101, 1988.
10. P. H. Zipkin, *Foundations of Inventory Management*, New York: McGraw Hill, 2000.
11. P. Glasserman and Sridhar Tayur, "Sensitivity analysis for base-stock levels in multiechelon production-inventory systems," *Management Science*, Vol. 41, No. 2, pp. 263–281, 1995.
12. S. C. Graves and S. P. Willems, "Optimizing strategic safety stock placement in supply chains," *Manufacturing and Service Operations Management*,

Vol. 2, No. 1, pp. 68–83, Winter 2000. (Erratum, M&SOM, Vol. 5, No. 2, pp. 176–177, Spring 2003.)

13. S. Willems, "A Tutorial on Strategic Safety Stock Placement in Supply Chains," in Proc. 2004 INFORMS Conference on OR/MS Practice (Cambridge, MA), pp. 276-283, 2004.
14. B. A. Ogunnaike and W. H. Ray, *Process Dynamics, Modeling, and Control*, New York: Oxford University Press, 1994.
15. C. E. Garcia, D. M. Prett, and M. Morari, "Model predictive control: theory and practice - a survey," *Automatica*, Vol. 25, No. 3, pp. 335–348, 1989.
16. S. Tzafestas, G. Kapsiotis, and E. Kyriannakis, "Model-based predictive control for generalized production planning problems," *Computers in Industry*, Vol. 34, No. 2, pp. 201–210, 1997.
17. E. Perea-Lopez, B. E. Ydstie, and I. E. Grossmann, "A model predictive control strategy for supply chain optimization," *Computers and Chemical Engineering*, Vol. 27, pp. 1201–1218, 2003.
18. M. W. Braun, D. E. Rivera, M. E. Flores. W. M. Carlyle, and K. G. Kempf, "A model predictive control framework for robust management of multi-product multi-echelon demand networks," *Annual Reviews in Control (Special Issue on Enterprise Integration and Networking)*, Vol. 27, No. 2, pp. 229–245, 2003.
19. M. W. Braun, D. E. Rivera, W. M. Carlyle, and K. G. Kempf, "Application of model predictive control to robust management of multi-echelon demand networks in semiconductor manufacturing," *Simulation: Transactions of the Society for Modeling and Simulation International*, Vol. 79, No. 3, pp. 139–156, March 2003.
20. W. Wang, D.E. Rivera, and K.G. Kempf, "Centralized model predictive control strategies for inventory management in semiconductor manufacturing supply chains," in *Proc. 2003 American Control Conference (Denver)*, pp. 585–590, 2003.
21. W. Wang, D. E. Rivera, K. G. Kempf, and K. D. Smith, "A model predictive control strategy for supply chain management in semiconductor manufacturing under uncertainty," e-Proc. AIChE Annual Meeting (San Francisco), Paper 446d, 2003.
22. J. H. Lee, M. Morari, and C. E. Garcia, "State-space interpretations of Model Predictive Control," *Automatica*, Vol. 30, No. 4, pp. 707–717, 1994.
23. F. D. Vargas-Villamil, D.E. Rivera, and K.G. Kempf, "A hierarchical approach to production control of re-entrant semiconductor manufacturing lines," *IEEE Transactions on Control Systems Technology*, Vol. 11, No. 4, pp. 578–87, July 2003.
24. S. Shirodkar, C. Arnold, K. Kempf, and J. Fowler, "Modeling and simulating supply chains for increased performance and profitability," in *2000 Proc. Inter. Conf. Modeling and Analysis of Semiconductor Manufacturing (Tempe, AZ)*, pp. 346–352.
25. K. Kempf, K. Knutson, J. Fowler, B. Armbruster, P. Babu, and B. Duarte, "Fast accurate simulation of physical flows in demand networks," in *2001 Proc. Semiconductor Manufacturing Operational Modeling and Simulation*

Symposium (Seattle), pp. 111–116.
26. D. Helbing, "Traffic and related self-driven many particle systems," *Reviews of Modern Physics*, Vol. 73, pp. 1067–1141, 2001.
27. D. Marthaler, D. Armbruster, and C. Ringhofer, "A mesoscopic approach to the simulation of semiconductor supply chains," *Simulation: Transactions of the Society for Modeling and Simulation International*, Vol. 79, No. 3, pp. 157–163, 2003.
28. D. Armbruster, D. Marthaler, C. Ringhofer, K. Kempf, and T. C. Jo, "A continuum model for a re-entrant factory," *Operations Research*, submitted for publication.
29. D. Armbruster, P. Degond, and C. Ringhofer, "Continuum models for interacting machines," this volume.
30. E. Lefeber, R. A. van den Berg, and J. E. Rooda, "Modelling manufacturing systems for control: a validation study", this volume.
31. G. W. Godding and K. G. Kempf, "A modular, scalable approach to modeling and analysis of semiconductor manufacturing supply chains", *Proc. IV SIMPOI/POMS (Sao Paulo)*, p. 1000–1007, 2001.
32. G. W. Godding, H. S. Sarjoughian, and K. G. Kempf, "Semiconductor supply network simulation," in *2003 Proc. IEEE Winter Simulation Conf. (New Orleans)*, pp. 1593–1601, 2003.

CHAPTER 4

MODELLING MANUFACTURING SYSTEMS FOR CONTROL: A VALIDATION STUDY

Erjen Lefeber, Roel van den Berg, J.E. Rooda

Department of Mechanical Engineering, Technische Universiteit Eindhoven
P.O. Box 513, NL-5600 MB Eindhoven, The Netherlands
E-mail: [A.A.J.Lefeber,R.A.v.d.Berg,J.E.Rooda]@tue.nl

This contribution deals with the modelling of manufacturing systems for control. First the concept of effective process times is introduced as a means to arrive at relatively simple discrete event models of manufacturing systems based on measured data. Secondly, a control framework is presented. Thirdly, a validation study is presented which shows that the currently available PDE-models for describing manufacturing systems need further improvement. Finally some criteria are specified which a PDE-model should at least meet in order to be considered valid.

4.1. Introduction

The dynamics of manufacturing systems has been a subject of study for several decades[9,12]. Over the last years, manufacturing systems have become more and more complex. A good understanding of the dynamics of manufacturing systems has therefore become even more important.

A living cell can also be considered as a tiny manufacturing system which produces certain parts via a system of "protein machines" (enzyme molecules). Parts produced by one "machine" then move to other "machines" to be processed. For a better understanding of this cell-dynamics, experiences from studying the dynamics of manufacturing systems might be helpful, and vice versa.

The goal of this contribution is to introduce the "outsider" to recent developments in the modelling and control of manufacturing systems and to provide some references that can be used as starting points for the interested reader. Since no familiarity with manufacturing systems is assumed, in Section 4.2 first some terminology and basic properties of manufacturing systems are introduced.

A commonly accepted approach for modelling the dynamics of manufacturing systems is by means of discrete event models, in which each product and each individual production step is modelled in great detail. In Section 4.3, the concept of effective process times is introduced as a means to arrive at relatively simple discrete event models of manufacturing systems. When building a discrete event model, one usually tries to include all possible disturbances due to machine failures, availability of operators and tools, maintenance, breaks, etc. Instead of using this white-box approach, a workstation is considered to be a black-box, whose input-output behaviour can be determined from real manufacturing data.

Using effective process times as a means to arrive at a relatively simple discrete event model of a manufacturing system is only the first step in our framework to control a manufacturing system. This framework is presented in Section 4.4. A second important ingredient of this control framework is the accurate approximation of the discrete event model's dynamics by a continuous model. Recently, a new class of continuous models has been proposed to capture the dynamics of manufacturing systems. This new class of models (PDE-models) is introduced in Section 4.5.

In Section 4.6 a summary is given of PDE-models that have been proposed in literature so far. The dynamic behaviour resulting from these models is compared with the dynamic behaviour that results from discrete event simulation. Unfortunately, none of the presented models describes the dynamics satisfactorily. Since this validation study calls for improved models, Section 4.7 concludes this chapter with a list of elementary properties that valid models should satisfy.

4.2. Preliminaries

We first need to introduce a few basic quantities and the main principles for manufacturing system analysis. The items produced by a manufacturing system are called *lots*. Also the words product and job are commonly used. The total number of lots in a manufacturing system is called *wip* (work-in-process) w. To characterise the behaviour of a manufacturing system two important measures are being used. The first measure is the *throughput* δ, i.e., the number of lots per time-unit that leaves the manufacturing system. The second measure is the *flow time* φ, i.e., the time from release of a lot in the system until the finished lot leaves the system. Instead of flow time the words cycle time and throughput time are also commonly used.

Ideally, a manufacturing system should both have a high throughput and a low flow time or low wip. Unfortunately, these goals can not both be met simultaneously. These two goals are conflicting, as can be seen from Fig. 4.1.

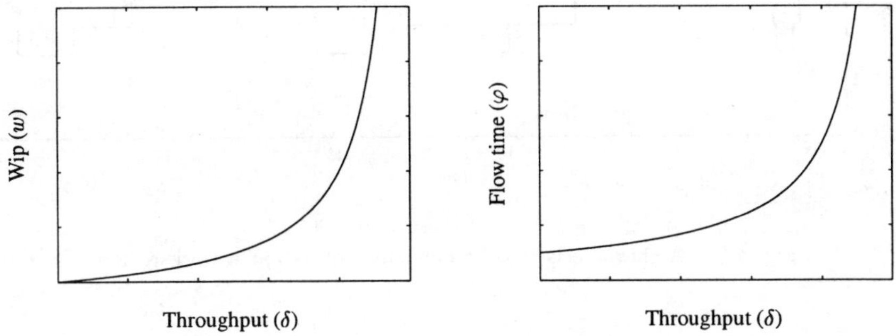

Fig. 4.1. Basic relations between basic quantities for manufacturing systems.

On the one hand, if we want to have a high throughput, we need to make sure that machines are always busy. Since from time to time disturbances like machine failures happen, we should make sure that we have buffers between two consecutive machines to make sure that the second machine can still continue if the first machine fails (or vice versa). As a result, for a high throughput we need to have many lots in the manufacturing system, i.e., we have a high wip. Therefore, if a new lot starts in the system it has a large flow time, since all lots that are currently in the system need to be completed first.

On the other hand, the least possible flow time can be achieved if a lot arrives at a completely empty system and never has to wait before processing at any machine takes place. As a result, for that system we have a small wip level, but also most of the time machines are not processing, yielding a small throughput.

When trying to control manufacturing systems, we need to make a trade-off between throughput and flow time, so the nonlinear (steady state) relations depicted in Fig. 4.1 need to be incorporated in any reasonable model of manufacturing systems.

Typical models of manufacturing systems are so-called discrete event models. In Fig. 4.2 we can see a characteristic graph of the wip at a workstation

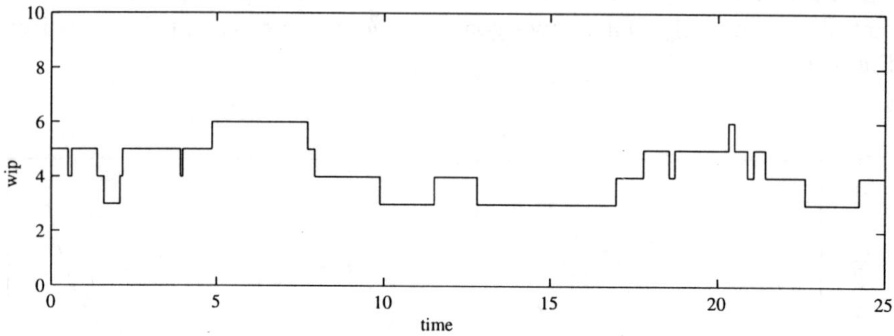

Fig. 4.2. A characteristic time-behaviour of wip at a workstation.

as a function of time. Wip always takes integer values with arbitrary (non-negative real) duration. One could consider a manufacturing system to be a system that takes values from a finite set of states and jumps from one state to the other as time evolves. This jump from one state to the other is called *event*. As we have a countable (discrete) number of states, the name of this class of models is explained.

The way we usually model a manufacturing system, is as a network of concurrent processes. For example, a buffer is modelled as a process that as long as it can store something is willing to receive new products, and as long as it has something stored is willing to send products. A basic machine is modelled as a process that wants to receive a product, delays for the period of processing and tries to send the product and keeps on doing these three consecutive things. The delay used is often a sample from some distribution.

In particular in the design phase discrete event models are used. These discrete event models usually contain a detailed description of everything that happens in the manufacturing system under consideration, resulting into large models. Since in practice manufacturing systems are changing continuously, it is very hard to keep these discrete event models up-to-date.

In the remainder of this chapter we introduce the concept of effective process times for arriving at simpler discrete event models than generally used. Next, we explain the control framework used for controlling manufacturing systems. In this framework, a crucial role is played by continuous approximation models of discrete event models. As these continuous approximation models should be valid, some validation studies are presented.

4.3. Effective Process Times (EPT's)

For the processing of a lot at a machine, many steps may be required. For example, it could be that an operator needs to get the lot from a storage device, setup a specific tool that is required for processing the lot, put the lot on an available machine, start a specific program for processing the lot, wait until this processing has finished (meanwhile doing something else), inspect the lot to determine if all went well, possibly perform some additional processing (e.g., rework), remove the lot from the machine and put it on another storage device and transport it to the next machine. At all of these steps something might go wrong: the operator might not be available, after setting up the machine the operator finds out that the required recipe can not be run on this machine, the machine might fail during processing, no storage device is available anymore so the machine can not be unloaded and is blocked, etc.

It is impossible to measure all sources of variability that might occur in a manufacturing system. One could incorporate some of these sources in a discrete event model. The number of operators and tools can be modelled explicitly and it is common practice to collect data on mean times to failure and mean times to repair of machines. Also schedules for (preventive) maintenance can be incorporated explicitly in a discrete event model. Nevertheless, still not all sources of variability are included. This is clearly illustrated in Fig. 4.3, obtained from[13]. The left graph contains actual realisations of flow times of lots leaving a real manufacturing system, whereas the right graph contains the results of a detailed deterministic simulation model and the graph in the middle contains the results of a similar model including stochasticity. It turns out that in reality flow times are much higher and much more irregular than simulation predicts. So, even if one tries hard to capture all variability present in a manufacturing system, still the outcome predicted by the model is far from reality.

Hopp and Spearman[12] use the term *effective process time* (EPT) as the time seen by lots from a logistical point of view. In order to determine this effective process time, Hopp and Spearman assume that the contribution of the individual sources of variability is known.

A similar description is given by Sattler[18] who defines the effective process time as all flow time except waiting for another lot. It includes waiting

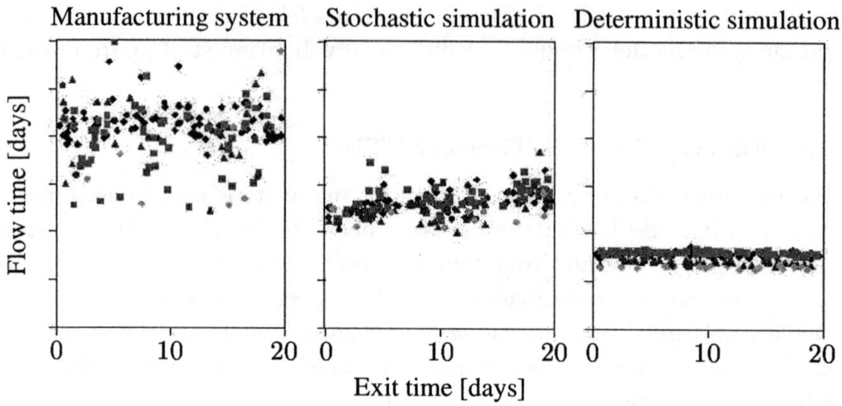

Fig. 4.3. A comparison.

for machine down time and operator availability and a variety of other activities. Sattler[18] noticed that her definition of effective process time is difficult to measure.

Instead of taking the bottom-up view of Hopp and Spearman, a top-down approach can also be taken, as shown by Jacobs et al.[13], where algorithms have been introduced that enable determination of effective process time realisations from a list of events. For these algorithms, the basic idea of the effective process time to include time losses was used as a starting point.

Consider the Gantt chart of Fig. 4.4. At $t = 0$ the first lot arrives at

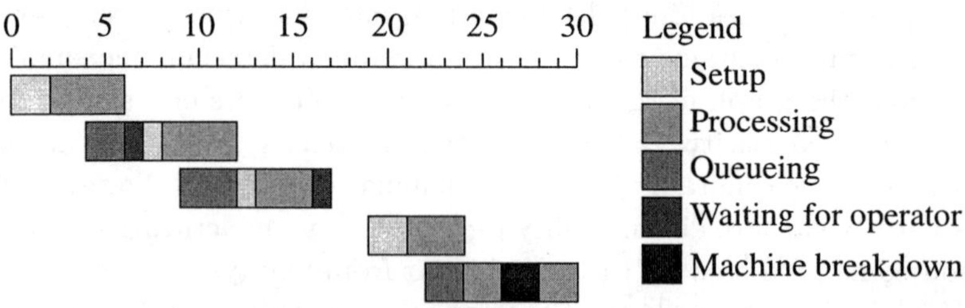

Fig. 4.4. Gantt chart of 5 lots at a workstation.

the workstation. After a setup, the processing of the lot starts at $t = 2$ and is completed at $t = 6$. At $t = 4$ the second lot arrives at the workstation.

At $t = 6$ this lot could have been started, but apparently there was no operator available, so only at $t = 7$ the setup for this lot starts. Eventually, at $t = 8$ the processing of the lot starts and is completed at $t = 12$. The fifth lot arrives at the workstation at $t = 22$, processing starts at $t = 24$, but at $t = 26$ the machine breaks down. It takes until $t = 28$ before the machine has been repaired and the processing of the fifth lot continues. The processing of the fifth lot is completed at $t = 30$.

If we take the point of view of a lot, what does a lot see from a logistical point of view? The first lot arrives at an empty system at $t = 0$ and departs from this system at $t = 6$. From the point of view of this lot, its processing took 6 time-units. The second lot arrives at a non-empty system at $t = 4$. Clearly, this lot needs to wait. However, at $t = 6$, if we would forget about the second lot, the system becomes empty again. So from $t = 6$ on there is no need for the second lot to wait. At $t = 12$ the second lot leaves the system, so from the point of view of this lot, its processing took from $t = 6$ till $t = 12$; the lot does not know whether waiting for an operator and a setup is part of its processing. Similarly, the third lot sees no need for waiting after $t = 12$ and leaves the system at $t = 17$, so it assumes to have been processed from $t = 12$ till $t = 17$. Following this reasoning, the resulting effective process times for lots are as depicted in Fig. 4.5.

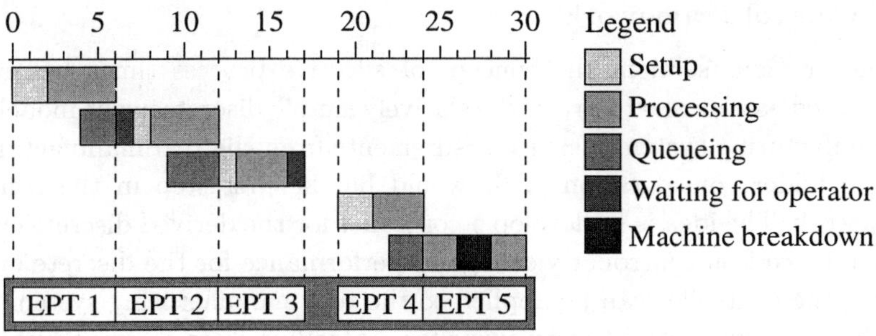

Fig. 4.5. EPT realisations of 5 lots at a workstation.

Notice that only arrival and departure events of lots to a workstation are needed for determining the effective process times. Furthermore, none of the contributing disturbances needs to be measured.

In highly automated manufacturing systems, arrival and departure events of lots are being registered, so for these manufacturing systems, effective process time realisations can be determined rather easily. Next, these

EPT realisations can be used in a relatively simple discrete event model of the manufacturing system. This discrete event model only contains the architecture of the manufacturing system, buffers and machines. The process times of these machines are samples from their EPT-distribution as measured from real manufacturing data. There is no need for incorporating machine failures, operators, etc., as this is all included in the EPT-distributions. Furthermore, the algorithms as provided in[13] are utilisation independent. That is, data collected at a certain throughput rate is also valid for different throughput rates. Also, machines with the same EPT-distribution can be added to a workstation. This makes it possible to study how the manufacturing systems responds in case a new machine is added, or all kinds of other what-if-scenario's. Finally, since EPT-realisations characterise operational time variability, they can be used for performance measuring. For more on this issue, the interested reader is referred to[4,13]. What is most important in the current setting, is that EPT's can be determined from real manufacturing data and yield relatively simple discrete event models of the manufacturing system under consideration. These relatively simple discrete event models serve as a starting point for controlling manufacturing systems.

4.4. Control Framework

In the previous section, the concept of effective process times has been introduced as a means to arrive at relatively simple discrete event models of a manufacturing system, using measurements from the real manufacturing system under consideration. This would be the first step in the control framework. The idea is to develop a controller for the derived discrete event model. Once this controller yields good performance for the discrete event model, the controller can be applied to the real manufacturing system.

Even though control theory exists for controlling discrete event systems, unfortunately none of it is appropriate for controlling real-life discrete event models of manufacturing systems. This is mainly due to the large number of states a manufacturing system can be in. Therefore, a different approach is needed.

If we concentrate on mass production, the distinction between lots is not really necessary and lots can be viewed in a more continuous way. Instead of the discrete event model we might consider an approximation model. This would be the second step in the control framework. Next, we can use standard control theory for deriving a controller for the approximation model.

These first three steps in the control framework are illustrated in Fig. 4.6.

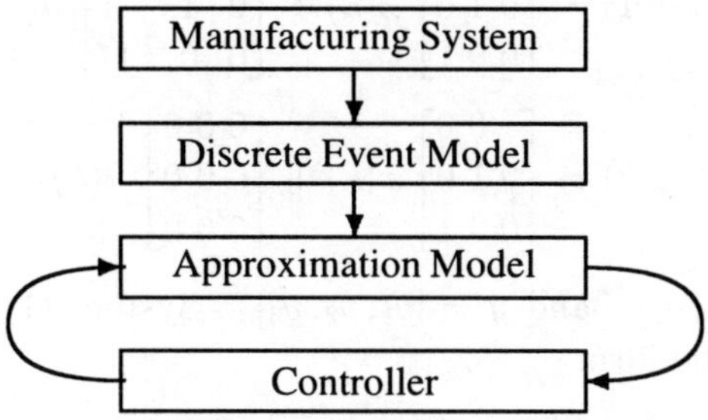

Fig. 4.6. Control framework (first three steps).

To make the second and third step more clear, a possible approximation model is presented in the next subsection, followed by a possible controller design based on this model. The final steps of the control framework conclude this section.

4.4.1. *Approximation Model*

Consider the manufacturing line in Fig. 4.7 which consists of two machines in series. Let $u_0(k)$ denote the number of lots that arrive at the system during shift k, let $u_i(k)$ denote the number of lots which machine M_i produces during shift k, let $x_i(k) = y_i(k)$ denote the number of lots in buffer B_i at the beginning of shift k ($i \in \{1, 2\}$), and let $x_3(k) = y_3(k)$ denote the number of lots produced by the manufacturing system during shift k. Assume that machines M_1 and M_2 have a maximum capacity of respectively μ_1 and

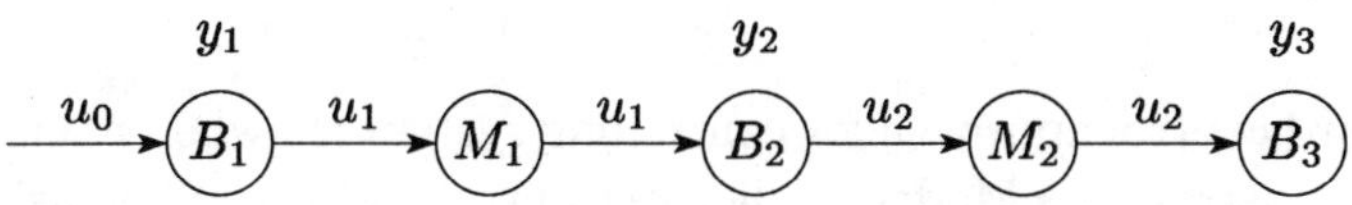

Fig. 4.7. A manufacturing line.

μ_2 lots per shift. Then we obtain the following approximation model:

$$x(k+1) = \begin{bmatrix} 1 & 0 & 0 \\ 0 & 1 & 0 \\ 0 & 0 & 1 \end{bmatrix} x(k) + \begin{bmatrix} 1 & -1 & 0 \\ 0 & 1 & -1 \\ 0 & 0 & 1 \end{bmatrix} u(k), \tag{1a}$$

$$y(k) = \begin{bmatrix} 1 & 0 & 0 \\ 0 & 1 & 0 \\ 0 & 0 & 1 \end{bmatrix} x(k) + \begin{bmatrix} 0 & 0 & 0 \\ 0 & 0 & 0 \\ 0 & 0 & 0 \end{bmatrix} u(k), \tag{1b}$$

where $u = [u_0, u_1, u_2]^\top$ and $y = [y_1, y_2, y_3]^\top$. System (1) is a controllable linear system of the form

$$\begin{aligned} x(k+1) &= Ax(k) + Bu(k), \\ y(k) &= Cx(k) + Du(k) \end{aligned}$$

as extensively studied in control theory (which explains the introduction of both x and y when deriving (1)). Therefore, many standard techniques from control theory can be used for deriving a controller for system (1).

4.4.2. *Model Predictive Control (MPC)*

For the continuous approximation model as derived in the previous section, we also have constraints. To be more precise, we have capacity constraints on the input u, as well as constraints on the state x and output y (the buffer contents should remain positive). These constraints can be expressed by means of the following equations:

$$\begin{aligned} 0 \leq u_1(k) \leq \mu_1, \quad & x_1(k) \geq 0, \quad y_1(k) \geq 0, \\ 0 \leq u_2(k) \leq \mu_2, \quad & x_2(k) \geq 0, \quad y_2(k) \geq 0, \\ & x_3(k) \geq 0, \quad y_3(k) \geq 0. \end{aligned} \tag{2}$$

A standard control approach for controlling system (1) when having to deal with the constraints is Model Predictive Control (MPC). When using MPC, it is common practice to define a reference output $y_r(k)$ that the system (1) should track.

Assume that the buffer contents at the end of shift $k-1$ have just been measured,i.e., $y(k)$. Since $y(k) = x(k)$, the current state of the system is known. So using this measurement, for each possible plan of future inputs $u(k), u(k+1), \ldots, u(k+p-1)$, by means of model (1) the resulting future outputs $y(k+1), y(k+2), \ldots, y(k+p)$ can be determined, usually denoted as $y(k+1|k), \ldots, y(k+p|k)$ to illustrate that these are predictions, while currently being at time-instant k. Next, costs can be associated with each possible plan of future inputs. Usually, these costs consist of both a penalty for not being exactly at the desired reference output, and a penalty for the control effort used. In this way almost naturally an optimisation problem arises, as the expected costs should be minimised over all possible plans for future inputs. A typical optimisation problem using this approach would be:

$$\min_{u(k),\dots,u(k+p-1)} \sum_{i=1}^{p} [y(k+i|k) - y_r(k+i)]^\top Q[y(k+i|k) - y_r(k+i)] + u(k+i-1)^\top Ru(k+i-1) \tag{3a}$$

subject to

$$0 \le u_j(k+i-1) \le \mu_j, \quad j \in \{1,2\}, \quad y(k+i|k) \ge 0 \quad (i = 1,\dots,p). \tag{3b}$$

As all $y(k+i|k)$ are affine functions of $u(k+i-1)$, this optimisation problem is a quadratic program which can be solved easily. From the resulting optimal solution $u^*(k),\dots,u^*(k+p-1)$, only $u^*(k)$ is used as the production targets for the next period. Since disturbances might occur, this optimisation procedure is repeated each shift, resulting in a receding horizon scheme.

The mentioned scheme is one of the simplest versions of MPC. The interested reader is referred to literature, e.g., [5,7,17] for more information about this control strategy.

Having illustrated the second and third step of the control framework, the final steps of the framework can be introduced.

4.4.3. *Control Framework (revisited)*

At the beginning of this section the first steps of the control framework have been explained, cf. Fig. 4.6. Using effective process times a relatively simple discrete event model of a manufacturing system can be derived based on measurements from the real manufacturing system. Next, an approximation model of the discrete event model can be derived. Subsequently, by means of standard control theory a controller for this approximation model can be derived.

When the derived controller behaves as desired, as a fourth step this controller could be connected to the discrete event model. This can not be done straightforwardly, since the derived controller is not a discrete event controller. The control actions still need to be transformed into events. It might very well be that the optimal control action would be to produce

2.75 lots during the next shift. One still needs to decide how many jobs to really start (2 or 3), and also when to start them. This is the left conversion block in Fig. 4.8. From this figure, it can also be seen that a conversion is

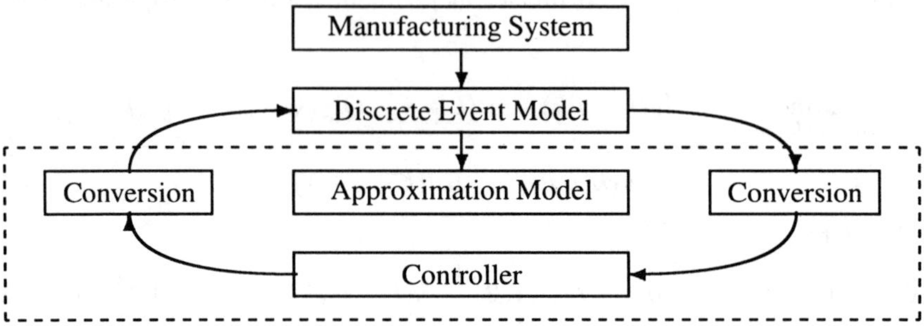

Fig. 4.8. Control framework (fourth step).

needed from discrete event model to controller. In the example treated in this section, it was decided to sample the discrete event model once every shift. Other strategies might be followed. For example, if at the beginning of a shift a machine breaks down it might not be such a good idea to wait until the end of the shift before setting new production targets. Designing proper conversion blocks would be the fourth step in the control framework.

After the fourth step, i.e., properly designing the two conversion blocks, a suitable discrete event controller for the discrete event model is obtained, as illustrated in Fig. 4.8 (dashed).

Eventually, as a fifth and final step, the designed controller can be disconnected from the discrete event model, and attached to the manufacturing system.

In the presented control framework two crucial steps can be distinguished. First, the discrete event model should be a good enough approximation of the real manufacturing system. For that reason, once a discrete event model of a manufacturing system has been made, the model needs to be validated. If results as shown in Fig. 4.3 are obtained the model needs further improvement. Second, the approximation model should be a good enough approximation of the discrete event model, or actually: of the discrete event model and the conversion blocks, since that is the system that needs to be controlled by the continuous controller. For that reason, approximation models of (discrete event models of) manufacturing systems also need to be validated. In the remainder of this chapter, some validation studies of approximation models are presented.

4.5. Modelling Manufacturing Systems

In the previous section, a control framework for controlling a manufacturing system has been presented. Similar ideas can be applied to the problem of controlling a network of interacting manufacturing systems. An illustrative example of a semiconductor manufacturing supply chain is given in Fig. 4.9.

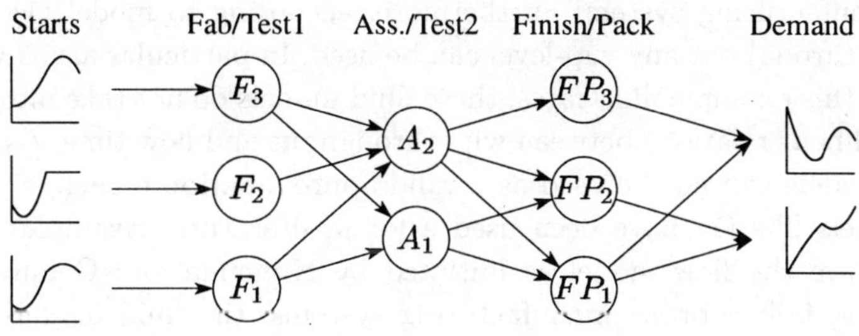

Fig. 4.9. A small semiconductor manufacturing supply chain.

In this figure, F_1, F_2, and F_3 denote wafer fabs, in which wafers are being produced, containing hundreds to thousands of integrated circuits (ICs) on its surface. Due to, among others, the large number of process steps, the re-entrant nature of the process flow, and the advanced process technologies, the fabrication of wafers is a complex manufacturing process. A typical flow time for a wafer fab is in the order of two months. That is, once a bare silicon wafer enters the manufacturing system, it typically takes about two months for the wafer to be completed.

Finished wafers are moved to an Assembly/Test facility, where individual chips are cut out of the wafer and each separated IC is assembled. Typical flow times for the manufacturing systems A_1 and A_2 are in the order of ten days. Finally, the chips are packaged in FP_1, FP_2, FP_3, and can be shipped to customers. This takes in the order of five days.

The control of this supply chain is one of the problems the semiconductor industry currently faces. The fact that flow times are large and nonlinearly dependent on the load (cf. Fig. 4.1) is one of the most difficult aspects in this problem. Notice that, even though the flow time of a wafer fab is in the order of two months, the raw process time of a wafer is less than two weeks. That is, if a wafer enters an empty wafer fab it would take less than two weeks for the wafer to be completed. This illustrates that the nonlinear relations between wip, throughput and flow time should be present in any approximation model.

Consider the manufacturing line depicted in Fig. 4.7 and assume that we start with an initially empty system and then turn on the machines, i.e., assume that $x(0) = 0$ and $u(k) = [\lambda, \lambda, \lambda]^\top$, $(0 < \lambda \leq \min(\mu_1, \mu_2)$, $k = 1, 2, 3, \dots)$, model (1) predicts that products immediately start leaving the manufacturing system. Furthermore, according to model (1) for each feasible throughput any wip-level can be used. In particular also a wip-level of 0. As this example illustrates, these fluid models do not take into account the nonlinear relations between wip, throughput and flow time. As a result, these models can not be used as a valid approximation model.

Models like (1) have been used a lot in literature. Examples of these models are the flow model as initiated by Kimemia and Gershwin[14] for modelling failure-prone manufacturing systems, the fluid models or fluid queues as proposed by queueing theorists[10], or the stochastic fluid model as introduced by Cassandras et al.[8]

Recently, a new class of models for manufacturing systems has been introduced[2,3,16]. In these models, the flow of products through a manufacturing system is modelled in a similar way as the flow of cars on a highway. Not only is the number of lots assumed to be continuous, also the position of a lot in the manufacturing system is assumed to vary continuously.

Let $t \in \mathbb{R}^+$ denote the time and let $x \in [0, 1]$ denote the position of a lot in the manufacturing line (the degree of completion). The behaviour of lots flowing through the manufacturing line can be described by three variables that vary with time and position: *flow* $u(x, t)$, measured in unit lots per unit time, *density* $\rho(x, t)$, measured in unit lots per degree of completion, and *speed* $v(x, t)$, measured in degree of completion per unit time. First, we observe that flow is the product of density and speed:

$$u(x, t) = \rho(x, t)v(x, t). \tag{4}$$

Second, assuming no scrap, the number of products between any two "locations" x_1 and x_2 ($x_1 < x_2$) needs to be conserved at any time t, i.e., the change in the number of products between x_1 and x_2 is equal to the flow entering at x_1 minus the flow leaving at x_2:

$$\frac{\partial}{\partial t}\int_{x_1}^{x_2} \rho(x,t)\,\mathrm{d}x = u(x_1,t) - u(x_2,t),$$

or in differential form:

$$\frac{\partial \rho}{\partial t}(x,t) + \frac{\partial u}{\partial x}(x,t) = 0. \tag{5}$$

Relations (4) and (5) are basic relations that any model must satisfy. As we have three variables of interest, (at least) a third relation is needed. Several choices can be made for this third (or more) relation(s), as the next section and[1] make clear.

As far as we know, the PDE-models as just described are the only ones that are solvable in limited time, describe the dynamics of a manufacturing system and incorporate both throughput and flow time. Flow or fluid models, like the one presented in (1), do not incorporate the nonlinear relation between throughput and flow time. Discrete event models do incorporate the nonlinear relation between throughput and flow time, but simulating these models takes a lot of time, making their on-line use computationally infeasible. A third class of models are queueing models like in[6,19]. They provide many insights in steady state behaviour of manufacturing lines, but the dynamics of manufacturing lines is rarely addressed.

Even though discrete event simulation is computationally intensive and queueing theory is mainly concerned with steady state, results from these models can be used for validating PDE-models. A minimal requirement for a valid PDE-model would be that that its steady state behaviour is in accordance with results from queueing theory. Also the dynamics of a PDE-model should be in accordance with the dynamics obtained from discrete event simulation. These checks are discussed in the next section, where queueing theory and discrete event simulation are used to validate PDE-models.

4.6. Validation of PDE-Models

In the previous section we introduced PDE-models as a way to model manufacturing systems. We only mentioned the basic ingredients (4), (5), and the need for a third relation. Also, we mentioned that results from queueing theory and from discrete event simulation can be used for validating PDE-models.

In this section we present validation studies for five PDE-models for manufacturing systems, using queueing theory for deriving the proper steady state and discrete event simulation for validating the dynamics.

4.6.1. *Manufacturing Systems*

When we consider the supply chain in Fig. 4.9, two typical manufacturing systems can be considered. On the one hand we have the factories F_1, F_2, and F_3, which have a re-entrant nature, on the other hand we have the factories A_1, A_2, FP_1, FP_2, and FP_3, which have the nature of a line of workstations. Therefore, we define two manufacturing systems:

Manufacturing System 1 A line consisting of 15 identical workstations (see Fig. 4.10). Lots visit the workstations according to the following recipe: 1-2-3-4-5-6-7-8-9-10-11-12-13-14-15. This is an "ordinary" manufacturing line, cf. A_1, A_2, FP_1, FP_2, and FP_3.

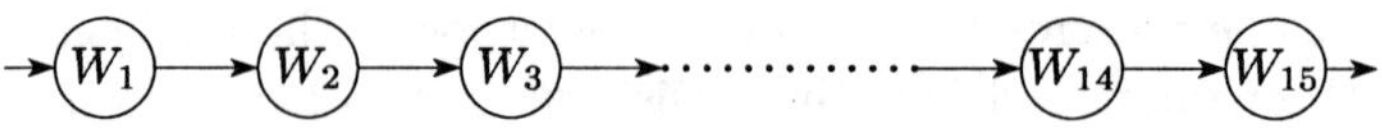

Fig. 4.10. An ordinary manufacturing line.

Manufacturing System 2 A line consisting of five identical workstations. Lots visit the workstations according to the following recipe: 1-2-3-4-5-1-2-3-4-5-1-2-3-4-5 (see Fig. 4.11). Since each lot re-enters the system twice, this is a re-entrant manufacturing line, cf. F_1, F_2, and F_3.

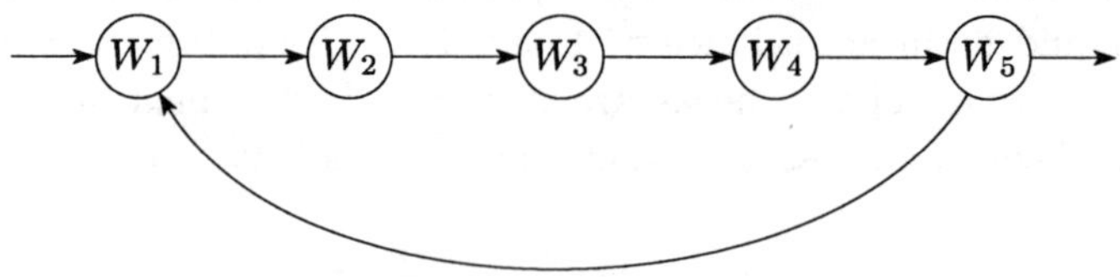

Fig. 4.11. A re-entrant manufacturing line.

We assume that each workstation consists of an infinite buffer, which operates under a FIFO policy (First In First Out), and a single machine whose effective process times are drawn from an exponential distribution with mean 1. If we furthermore assume that lots arrive at the manufacturing system according to a Poisson process with an arrival rate λ, we can derive the following steady state properties by means of queueing theory:

- For Manufacturing System 1 (Fig. 4.10), the mean number of lots equals $\frac{\lambda}{1-\lambda}$ in each workstation, resulting in a mean number of $\frac{15\lambda}{1-\lambda}$ lots in the system. Furthermore, the mean flow time of lots for Manufacturing System 1 is $\frac{15}{1-\lambda}$.
- For Manufacturing System 2 (Fig. 4.11), the mean number of lots equals $\frac{3\lambda}{1-3\lambda}$ in each workstation, resulting in a mean number of $\frac{15\lambda}{1-3\lambda}$ lots in the system. Furthermore, the mean flow time of lots for Manufacturing System 2 is $\frac{15}{1-3\lambda}$.

4.6.2. *PDE-Models*

In the validation studies we consider the following five models that have been proposed in literature.

Model 1: Single queue I[2] Relations (4), (5) together with

$$v(x,t) = \frac{\mu}{1 + \int_0^1 \rho(s,t)\,\mathrm{d}s}, \tag{6}$$

where $\mu > 0$ is a constant representing the processing rate of the workstation.

Model 2: Single queue II[2] Relations (4), (5) together with

$$\frac{\partial \rho v}{\partial t}(x,t) + \frac{\partial \rho v^2}{\partial x}(x,t) = 0, \tag{7}$$

and the additional boundary condition

$$\rho v^2(0,t) = \frac{\mu \cdot \rho v(0,t)}{1 + \int_0^1 \rho(s,t)\,\mathrm{d}s}, \tag{8}$$

where $\mu > 0$ again denotes the processing rate of the workstation.

Model 3: Re-entrant I[3] Relations (4), (5) together with

$$v(x,t) = v_0 \left(1 - \frac{1}{L_{\max}} \int_0^1 \rho(s,t)\,\mathrm{d}s\right), \tag{9}$$

where $v_0 > 0$ is a constant representing the maximal speed that can be achieved (i.e., $1/v_0$ denotes the theoretical minimal flow time), and $L_{\text{max}} > 0$ is a constant representing the maximal number of lots that is allowed in the manufacturing system.

Model 4: Re-entrant II[3] Relations (4), (5) together with (7), and the additional boundary condition

$$\rho v^2(0,t) = v_0 \left(1 - \frac{1}{L_{\text{max}}} \int_0^1 \rho(s,t)\,\mathrm{d}s\right) \rho v(0,t), \tag{10}$$

where v_0 and L_{max} are the same as in (9).

Model 5: *m* identical machines[16] Relations (4), (5) together with

$$v(x,t) = \frac{\mu}{m + \rho(x,t)}, \tag{11}$$

where $m > 0$ denotes the number of machines, and $\mu > 0$ denotes the processing rate of each workstation.

All five models have the boundary condition

$$\rho v(0,t) = \lambda(t),$$

where $\lambda(t)$ denotes the inflow to the manufacturing system in unit lots per unit time.

Recently, other PDE-models than models 1–5 have been proposed, cf. [1] (an other chapter in this book). These models have not been incorporated in these validation studies.

4.6.3. *Validation Study*

In the previous subsections we introduced Manufacturing Systems 1 and 2, as well as the PDE-models used in the validation studies.

Manufacturing System 1 is an ordinary manufacturing line and does not have a re-entrant nature. As PDE-models 3 and 4 have been specifically designed for re-entrant manufacturing systems, they have not been used in the validation studies for Manufacturing System 1.

As mentioned, according to queueing theory the mean number of lots in Manufacturing System 1 is $\frac{15\lambda}{1-\lambda}$ in case we have a mean arrival rate of λ. Translated into PDE-terms we have in equilibrium

$$\rho(x,t) = \frac{15\lambda}{1-\lambda}, \qquad v(x,t) = \frac{1-\lambda}{15}. \tag{12}$$

From (12) we obtain, by eliminating λ, that in steady state

$$v(x,t) = \frac{1}{15 + \rho(x,t)} = \frac{1}{15 + \int_0^1 \rho(s,t)\,\mathrm{d}s}. \tag{13}$$

From (13) we obtain that the models 1, 2, and 5 are valid in steady state, provided that in (6) and (8) we replace the denominator $1+\int_0^1 \rho(s,t)\,\mathrm{d}s$ with $15+\int_0^1 \rho(s,t)\,\mathrm{d}s$, which is consistent with the results in[2]. In[2] a single queue is assumed. If, instead, we assume a line of 15 workstations the mentioned modification of (6) and (8) results.

For Manufacturing System 2, the mean number of lots in the system equals $\frac{15\lambda}{1-3\lambda}$ with a mean flow time of $\frac{15}{1-3\lambda}$. Translated into PDE-terms we have in equilibrium

$$\rho(x,t) = \frac{15\lambda}{1-3\lambda}, \qquad v(x,t) = \frac{1-3\lambda}{15}. \tag{14}$$

From (14) we obtain, by eliminating λ, that in steady state

$$v(x,t) = \frac{1}{15+3\rho(x,t)} = \frac{1}{15}\left(1-\frac{\int_0^1 \rho(s,t)\,\mathrm{d}s}{\frac{5}{1-3\lambda}}\right). \tag{15}$$

When we compare (15) with (9) and (10), we notice that in order for models 3 and 4 to be valid in steady state, we need $L_{\max} = \frac{5}{1-3\lambda}$, where λ denotes the steady state arrival rate. Since $L_{\max}$ depends on λ, the re-entrant models 3 and 4 are not likely to be "globally" valid for re-entrant manufacturing systems, i.e., valid for an arbitrary arrival rate λ. In the best case they are valid "locally" around a certain λ. On the other hand, any manufacturing system can contain only a finite number of lots, arguing the validity of a queueing model with infinite buffers.

From (15) we obtain that the models 1, 2, and 5 are valid in steady state, provided that in (6) and (8) we replace the denominator $1+\int_0^1 \rho(s,t)\,\mathrm{d}s$ with $15+3\int_0^1 \rho(s,t)\,\mathrm{d}s$, and in (11) we replace the denominator with $15+3\rho(x,t)$. The former can be argued to be a suitable model for a re-entrant manufacturing line (a homogeneous velocity over the line is used), whereas the latter is not a proper model for a re-entrant manufacturing line. It would have been better to replace it with $15+\rho(x,t)+\rho(\langle x+\frac{1}{3}\rangle_1,t)+\rho(\langle x+\frac{2}{3}\rangle_1,t)$. where $\langle\alpha\rangle_1$ denotes α modulo 1, i.e., all digits behind the decimal separator.

Next, we can use discrete event models of System 1 and System 2 to study the dynamics of the proposed PDE-models. For this we used the specification language χ[11]. Starting with an initially empty system, we performed experiments where lots arrive according to a Poisson process with a mean arrival rate λ. During an experiment we collected at the times $t = 1, 2, 3, \ldots$ the number of lots in each workstation as well as the number of lots that has been completed by the system. In order to guarantee a 99% confidence interval with a relative width of less than 0.01 for *each* measurement, experiments have been repeated 1.000.000 times. We averaged all data, resulting in the average number of lots in each workstation, as well as the number of lots that has been completed by the system, at each time-instant. This we did for both Manufacturing System 1 and Manufacturing System 2, where we chose the arrival rate such that the steady state utilisation of the workstations was respectively 25%, 50%, 75%, 90%, and 95% (so $\lambda = 0.25, 0.5, 0.75, 0.90, 0.95$ for Manufacturing System 1 and $\lambda = 0.08333, 0.16667, 0.25, 0.3, 0.31667$ for Manufacturing System 2). These experiments provide more data than can be presented in this chapter. The interested reader is referred to[15] for more results. Here we present some general findings.

The first results we present are for Manufacturing System 1 with an arrival rate of $\lambda = 0.25$. Fig. 4.12 presents the evolution of the total number of lots in the system as a function of time. The solid line describes

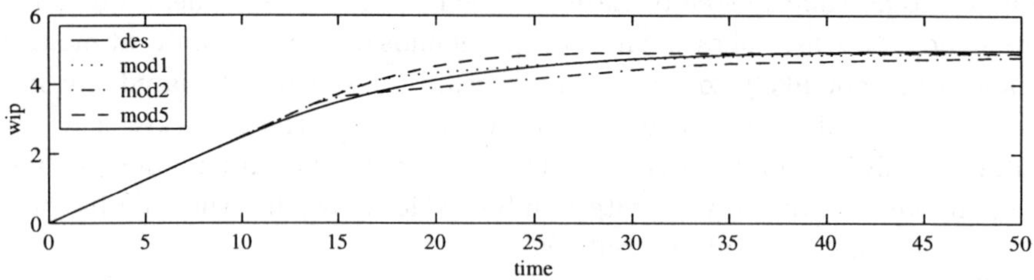

Fig. 4.12. Number of lots in Manufacturing System 1 for utilisation of 25%.

the (averaged) result of the discrete event simulations. The dotted line, the dash-dotted line, and the dashed line describe the result according to Model 1, Model 2, and Model 5 respectively. In Fig. 4.12 we see that initially the total number of lots in the line linearly increases. This is due to the fact that lots are only entering the system and it takes a while before lots start coming out. Also, we see that all models predict that in steady state five lots are in the system. This is as expected. When we closely look at Fig. 4.12 we see that around $t = 10$ the graph of the discrete event simulation bends off from the PDE-graphs, from which we can conclude that the moment at which the first lot leaves the system is overestimated by the PDE-models. That is, according to the discrete event simulation this should happen earlier. Also, we see that after $t = 40$ all three PDE-models underestimate the number of products in the system. Therefore, all PDE-models predict that the system is later in steady state than according to the discrete event simulation.

The differences in behaviour become more clear when we consider the development of the density over time. This can be made most clear by means of a movie, for which the reader is referred to[15]. In Fig. 4.13 the most important parts of the behaviour are captured. The figure presents respectively $\rho(0,t)$, $\rho(0.5,t)$ and $\rho(1,t)$, again for the discrete event model, Model 1, Model 2, and Model 5. For the discrete event system we assume the density to be piecewise constant at intervals of width $\frac{1}{15}$. When looking

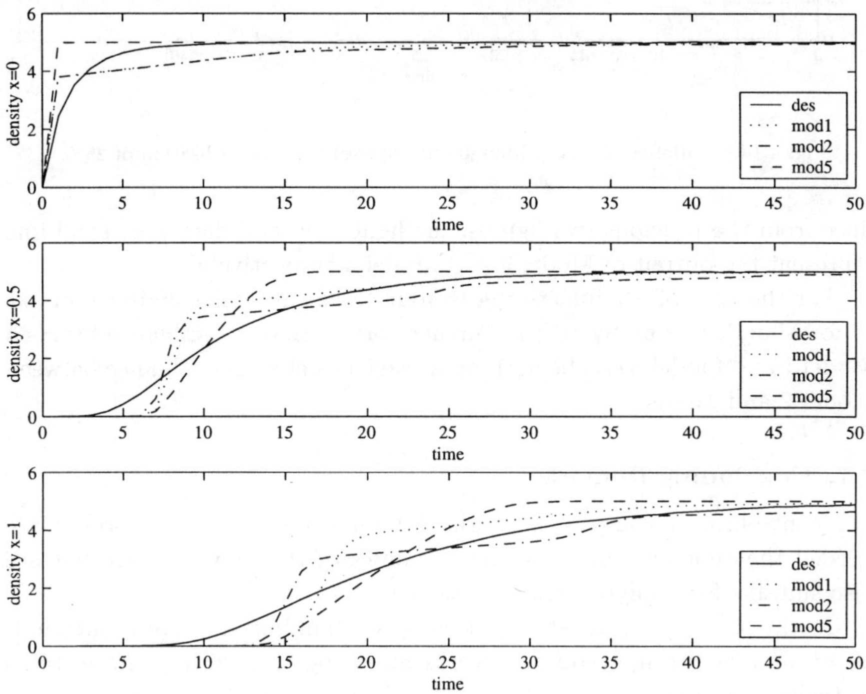

Fig. 4.13. Densities at $x = 0$, $x = 0.5$ and $x = 1$ for utilisation of 25%.

at the first graph, we see that the behaviour of Model 1 and Model 2 almost coincide. All three models predict a quicker raise of the density than the discrete event model predicts. If we look at the graph of $\rho(0.5,t)$ we see that initially the PDE-models underestimate the growth of the density, around

$t = 7$ the PDE-models show a strong increase in the density, resulting in an over-estimation of the density. Similar behaviour can be observed for $\rho(1, t)$.

The second results we present are for Manufacturing System 2 with an arrival rate of $\lambda = 0.08333$. Fig. 4.14 presents the evolution of the total number of lots in the system as a function of time. In addition to the

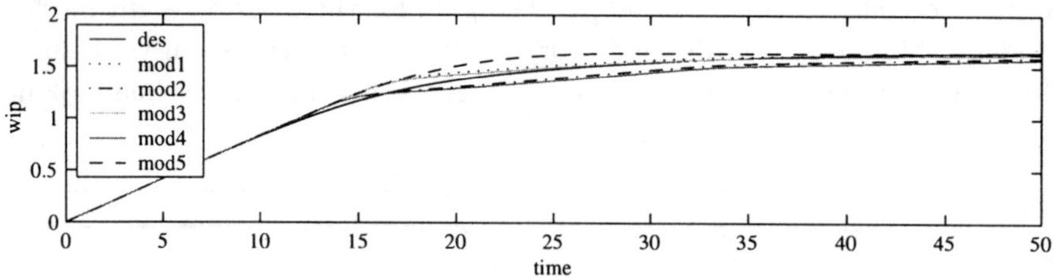

Fig. 4.14. Number of lots in Manufacturing System 2 for utilisation of 25%.

lines from the previous two figures, the light grey and dark grey solid line represent the output of Model 3 and Model 4 respectively.

For the case of Manufacturing System 2 we can make similar remarks as for Manufacturing System 1. Furthermore, a close resemblance between Model 1 and Model 3 can be noticed, as well as a close resemblance between Model 2 and Model 4.

4.7. Concluding Remarks

For controlling a complex network of interacting machines, models are needed that not only describe the dynamics of the network well, but are also suitable for applying control theory to.

In this chapter we illustrated that when building discrete event models of manufacturing systems, a workstation together with all its possible disturbances might be considered as a black box. Instead of using white box modelling and trying to capture all possible disturbances well (as is often done), one can also focus on accurately capturing the input-output behaviour. For this, the concept of effective process times (EPT's) can be used. Using this approach it is possible to arrive at valid discrete event models of manufacturing systems, using real manufacturing data.

Though EPT's can be used to arrive at valid discrete event models of manufacturing systems, these discrete event models can not be used for deriving suitable controllers. Therefore, a control framework has been presented which makes use of an approximation model of the discrete event model. PDE-models have been mentioned as possible approximation models that are solvable in limited time, describe the dynamics of a manufacturing system and incorporate both throughput and flow time. However, the question remains: are these PDE-models valid models of manufacturing systems?

The presented validation studies showed that more accurate PDE-models are needed. Recently, new models have been proposed (cf. [1]) which might do a better job. Nevertheless, most of the recently developed PDE-models fail at least at one of the following elementary tests:

- Given a fixed set of model parameters, the correct steady state wip is achieved for arbitrary constant influx λ. Models 3 and 4 fail this test.
- For a manufacturing line, as depicted in Fig. 4.10, lots at the end of the line are not influenced by lots in the beginning of the line. In particular, assume the system is in a certain steady state and suddenly the influx decreases. This is not immediately noticed in the outflux. Models 1 and 2 fail this test.
- For a manufacturing system the steady state wip distribution is not only determined by the influx (arrival rate of lots), but also by its variance.

  In the validation studies as presented, Poisson arrivals have been assumed, yielding a homogeneous wip distribution over the line (cf. (12) and (14)). However, if we would have used constant inter arrival times the wip distribution would not be homogeneous, but as depicted in Fig. 4.15. In case the inter arrival times would have

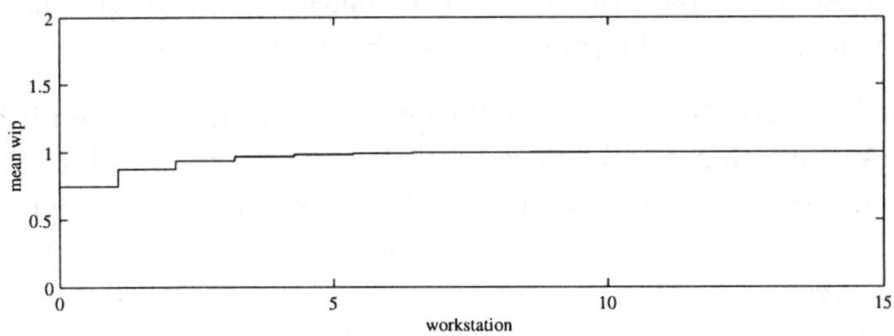

Fig. 4.15. Manufacturing System 1 with deterministic arrivals.

had a higher variance, the wip distribution as depicted in Fig. 4.16 might arise. For the same influx different steady state wip levels

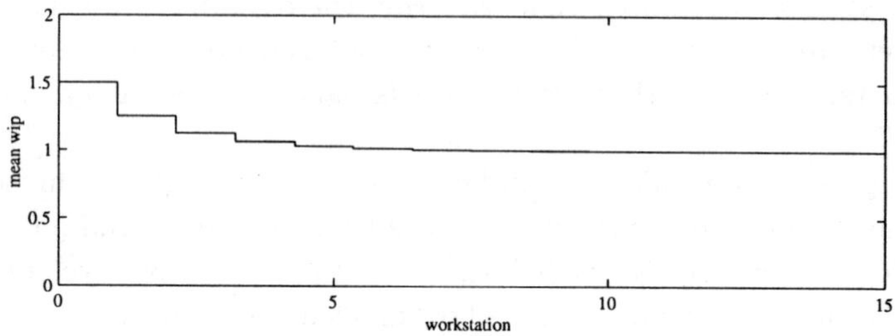

Fig. 4.16. Manufacturing System 1 with highly irregular arrivals.

results, depending on the variance of the influx. The higher the variance, the higher the steady state wip level.
Since the variance of the influx is not a system property, for a given set of model parameters, different steady state wip profiles should be achievable for different variances of the influx.
Models 1, 3, and 5 fail this test. With the current boundary conditions Models 2 and 4 fail this test too.

- In a manufacturing system, lots do not flow back wards. Assume that the first and the last machine of a manufacturing system fail. In that case both the influx and the outflux are zero. Furthermore, assume that initially the beginning of the line is empty (say $\rho(x,0) = 0$ for $0 \leq x \leq \frac{1}{2}$), but that the end of the line contains some lots. Assume that both the influx and the outflux remain zero. This should also hold for a valid PDE-model.
  Models 1–5 pass this test. Nevertheless, this is a test that models which do pass the previous test should also pass.

A final remark deals with the effects of correlation between influx and outflux. Assume that a to be developed PDE-model passes all of the above mentioned tests. If the presented control framework is used to derive a control strategy for the manufacturing system(s) under consideration, typically a feedback results. That is, the current state of the system determines what new influxes will be. This introduces correlations between the influx and the outflux.

To illustrate this, consider a manufacturing system consisting of only one workstation with an infinite buffer. Assume that process times are drawn from an exponential distribution with mean 1 and assume arrivals according to a Poisson process with a mean arrival rate of $\lambda = 0.5$. Using queueing theory the mean number of lots in this manufacturing system equals 1 and the mean flow time equals 2. Furthermore, the outflux is also a Poisson process with a mean departure rate of $\lambda = 0.5$. We also know from queueing theory that a Poisson arrival process with a mean arrival rate of $\lambda = 1$ results in an unstable system.

Now we apply feedback to this system. The policy we use is to keep the total number of products in this system equal to 1. For the resulting closed-loop system, the mean wip number of lots in the system also equals 1. However, the mean flow time becomes 1 and a mean throughput of 1 results. So the arrival and departure processes both are Poisson processes with a mean rate of 1, something which was unfeasible for the system without feedback.

The system itself has not changed. The only thing that changed was that in the former case the influx and outflux were uncorrelated, whereas in the latter case they were correlated. This example illustrates that apparently the possibility of correlation between influx and outflux should be incorporated in the models too.

The goal of this contribution was to introduce the "outsider" to recent developments in the modelling and control of manufacturing systems and to provide some references that can be used as starting points for the reader who has become more interested. Firstly, we presented the concept of effective processing times. Instead of trying to model what is going on exactly, we try to capture only the input-output behaviour as good as possible using real manufacturing data. Secondly, we presented a framework for controlling manufacturing systems. Thirdly, we showed that the currently available PDE-models that try to capture the dynamics of manufacturing systems at a macroscopic level need further improvement.

References

1. D. Armbruster, P. Degond, and C. Ringhofer. Continuum models for interacting machines. In D. Armbruster, K. Kaneko, and A. Mikhailov, eds., *Networks of interacting machines: production organization in complex industrial systems and biological cells*. World Scientific Publishing, Singapore, 2005. This book.
2. D. Armbruster, D. Marthaler, and C. Ringhofer. Kinetic and fluid model hi-

erarchies for supply chains. *SIAM Journal on Multiscale Modeling and Simulation*, vol. 2, no. 1, pp. 43–61, 2004.
3. D. Armbruster, D. Marthaler, C. Ringhofer, K. Kempf, and T.-C. Jo. A continuum model for a re-entrant factory, 2004. URL `http://math.la.asu.edu/~chris`, submitted to Operations Research.
4. P.P. v. Bakel, J.H. Jacobs, L.F.P. Etman, and J.E. Rooda. Quantifying variability of batching equipment using effective process times., 2004. Submitted to IEEE Transactions on Semiconductor Manufacturing.
5. R.R. Bitmead, M. Gevers, and V. Wertz. *Adaptive Optimal Control — The Thinking Man's GPC*. Prentice Hall, Englewood Cliffs, 1990.
6. J.A. Buzacott and J.G. Shantikumar. *Stochastic Models of Manufacturing Systems*. Prentice Hall, Englewood Cliffs, New Jersey, USA, 1993.
7. E.F. Camacho and C. Bordons. *Model Predictive Control*. Springer, London, 1999.
8. C.G. Cassandras, Y. Wardi, B. Melamed, G. Sun, and C. Panayiotou. Perturbation analysis for on-line control and optimization of stochastic fluid models. *IEEE Transactions on Automatic Control*, vol. 47, no. 8, pp. 1234–1248, 2002.
9. J.W. Forrester. *Industrial Dynamics*. MIT Press, Cambridge, Massachusetts, USA, 1961.
10. J.M. Harrison. *Brownian Motion and Stochastic Flow Systems*. John Wiley, New York, 1995.
11. A.T. Hofkamp and J.E. Rooda. Chi reference manual. Tech. rep., Technische Universiteit Eindhoven, Eindhoven, The Netherlands, 2002. URL `http://se.wtb.tue.nl/documentation/`.
12. W.J. Hopp and M.L. Spearman. *Factory Physics*. Irwin/McGraw-Hill, New York, USA, second edn., 2000.
13. J.H. Jacobs, L.F.P. Etman, E.J.J. v. Campen, and J.E. Rooda. Characterization of the operational time variability using effective processing times. *IEEE Transactions on Semiconductor Manufacturing*, vol. 16, no. 3, pp. 511–520, 2003.
14. J. Kimemia and S.B. Gershwin. An algorithm for the computer control of a flexible manufacturing system. *IIE Transactions*, vol. 15, no. 4, pp. 353–362, 1983.
15. E. Lefeber. Movies containing results of the validation studies, 2004. URL `http://se.wtb.tue.nl/~lefeber/berlinbook/movies/`.
16. E. Lefeber. Nonlinear models for control of manufacturing systems. In G. Radons and R. Neugebauer, eds., *Nonlinear Dynamics of Production Systems*, chap. 5, pp. 69–81. John Wiley, Berlin, Germany, 2004.
17. J.M. Maciejowski. *Predictive Control with Constraints*. Prentice Hall, 2000.
18. L. Sattler. Using queueing curve approximations in a fab to determine productivity improvements. In *Proceedings of the 1996 IEEE/SEMI Advanced Semiconductor Manufacturing Conference and Workshop*, pp. 140–145. Cambridge, MA, USA, 1996.
19. R. Suri. *Quick Response Manufacturing: A Companywide Approach to Reducing Lead Times*. Productivity Press, Portland, Oregon, USA, 1998.

CHAPTER 5

ADAPTIVE NETWORKS OF PRODUCTION PROCESSES

Adam Ponzi

Department of Physics and Related Technology,
The University of Palermo, Viale delle Scienze, Palermo, Italy
E-mail: ponzi@lagash.dft.unipa.it

A dynamical model of an adaptive production network is proposed as an extension of von Neumann's static model. Sets of input products are jointly converted into sets of output products by production processes. The rate of production of a given production process is limited by the minimum quantity of its set of input commodities. Switching between different choices of the minimum input commodity leads to the appearance of cyclic behaviour. The generation of complex oscillations with multiple timescales is also shown when several processes are combined to form a chain or a network. This model is primarily applied to understanding the non-equilibrium behaviour of a production economy but we also discuss its relevance to the understanding of some of the complex features of biological cellular organization and behaviour.

5.1. Introduction

A production economy and a living cell are united by the fact that they are both *adaptive networks of production processes, catalyzed by machines.* In our recent research we have made a model to describe such networks[1,2,3]. It is aimed at understanding how macro economics emerges from micro economics. In particular we wish to determine how complex non-equilibrium macro-economic behaviour such as economic production cycles, i.e. growth and recession, labour unemployment cycles, and price inflation and deflation emerge from the basic dynamics and evolution of adaptive networks of production processes which form the microscopic foundations of an economic production economy. It is fair to say that neoclassical economic theory, based on static and equilibrium analysis applied to idealized utility maximizing agents has totally failed to get to grips with the real nature of an economy.

While the main focus of this research is on economics, however, as I shall describe, it may also be relevant to the understanding of cellular biology as well. In particular adaptive networks of production processes may be useful for describing some of the complex features of cellular organization such as coherent activity of various parts of the cell, self-regulation, internal division of labour into specialized molecular machines, and the appearance of multiple-timescale internal cellular cycles.

Indeed a production economy, or a factory, is extraordinarily similar to a living cell[8,9,10,7]. In a cell, enzyme catalyst molecules behave like irreversible machines transforming input chemicals into output chemicals and forming complex molecular 'assembly lines'. Energy can become localized in ATP molecules and be transferred around between these molecular production processes, in a intriguingly similar way to money in an economy. Protein machines can be switched on an off by allosteric regulation, in a similar way to a factory manager regulating different production lines. Processes are generally irreversible and may be globally coupled since internal cellular diffusion times are much faster than the relaxation and folding dynamics of enzyme catalysts[7]. Of course economic processes are also irreversible and the free market mechanism provides a global coupling whereby products are valued and exchanged and some production processes become more active while others disappear. In both an economy and a living cell, this coherent complex organization is the outcome of a competitive evolutionary efficiency maximizing process.

This new model is based on a combination of von Neumann's neoclassical model of economic production (VNM)[5,6] and the general dynamics of catalytic reaction networks[21,22]. Our model is based on the VNM but is very different. As in the VNM our model represents production processes like chemical reactions where factors of production are considered conserved in the reaction like chemical catalysts, see Fig.5.1(a). However our model is a non-equilibrium dynamical model, described by a system of differential equations or a coupled mapping in its discrete time version. The analysis does not proceed by imposing optimization relationships such as perfect competition or profit maximization or cost minimization and then deriving the structure of an equilibrium as in typical analysis of the VNM. Rather the emphasis is on dynamic numerical simulation on a computer to see what actual behaviour emerges. For example profit maximization may be found to be an *emergent* characteristic itself.

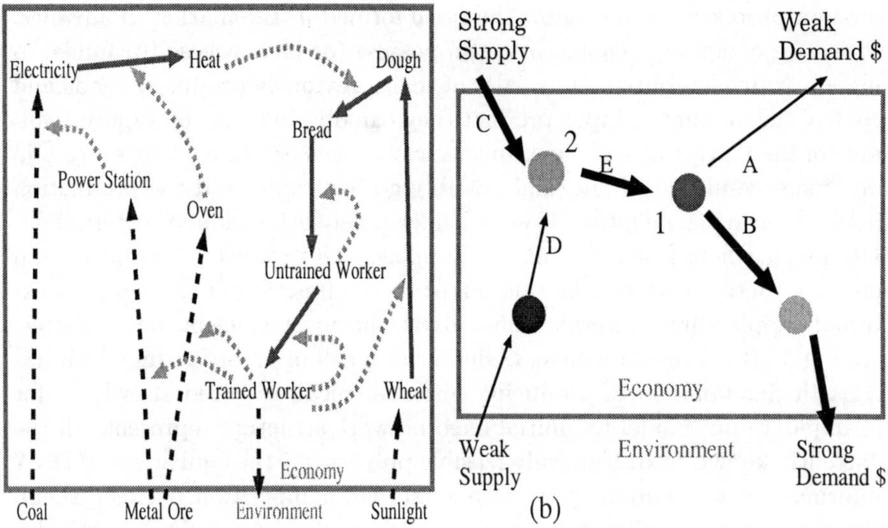

Fig. 5.1. (a) An VNM Economy describing the technologically feasible fixed set of production processes and products. Solid lines are production flows and dotted lines are catalytic effects. When many such processes are coupled together we obtain a production economy. Some processes are autocatalytic as shown. The system may be closed or open. In an open system some products are supplied from the environment and others are demanded, shown as dashed lines. The environment may of course be the set of other economies or other countries. (b) An adaptive network. Circles denote products transformed by production processes which are shown as arrows. In this simple example each process is assumed to only have one input product and one output product and catalysts are not shown for convenience. Product moves from external input supply to external output demand in the positive arrow direction. Funds (i.e. money) move from the external output demand to the external input supply in the negative arrow direction. Arrow size indicates the strength of the process, i.e. its funds size and therefore its share of the available input material. Processes A and B both demand product 1 and are therefore in competition for it. Process B has access to larger funds than process A so it takes the larger share. Similar processes C and D both supply product 2 and so are in competition for funds coming from process E. Process C makes product 2 from an external product in larger supply than process D does. Process C then takes a larger share of the available funds and has a larger size. In this way some pathways are *strengthened* while others may go *bankrupt* completely. The path from strong supply to strong demand forms a production chain, or an assembly line. In this way the network *adapts* or internally evolves to fulfill the *function* of changing its external supplies into external demands in the most efficient way possible, and the resulting network has an dynamically changing emergent structure.

In this model processes are globally coupled through a central market where they compete for goods they require and submit the goods they wish to sell. The product price is formed by market clearing condition and production processes do not know the price formed in the market in advance. Processes compete because they also possess (or have access to) funds, in money, which are obtained by sales of their previously produced goods and control the amount of input product they can obtain from the supply available in the market at any time. In this way processes can become successful and increase in size or fail (and possibly go bankrupt) and the production network becomes *adaptive.* If we consider a network inside an external environment where raw materials are supplied and products demanded from outside, the network can be said adapt to the function of changing its external supplies into external demands in the most efficient way possible, see Fig.5.1(b). This, of course, is similar to a cell in an organism which has a particular function of producing some chemical, required elsewhere, for example. In our model the initial fixed network structure represents all the currently known technologically feasible processes, - the equivalent of DNA information, while adaptation occurs through competition between them. Processes going bankrupt are technologies inefficient within a particular emergent network structure.

Furthermore since this model is fundamentally non-equilibrium, as in reality, processes need not always obtain supplies from the market in the correct stochiometric proportions they require for processing. They may obtain excess *unemployed* catalyst for example. To model this we impose a Leontief type production function, whereby the rate of processing is given by the *minimum* product supply possessed by the process. This is in contrast to a form such as the Law of Mass Action[23] where the processing rate depends on the concentrations of all the input products. Indeed such a production function is what one would expect to be appropriate for describing the kinetics of highly tuned and specialized *machines* rather than randomly colliding chemicals. This production function is of course also appropriate for describing cellular reaction kinetics, as noted above, and in fact is very similar in form to a threshold Michaelis-Menten reaction kinetics.

We now review the background of the von-Neumann model of production in more detail and then give the mathematical description of our new dynamical version.

5.2. Review of von-Neumann Model

In von-Neumann's model[24,25] of economic production each good is produced jointly with certain others like a chemical reaction. A process of production converts one bundle of goods, including capital equipment, into another bundle of goods, including the capital equipment. Capital goods are therefore treated something like catalysts in a chemical reaction, reformed at the end of the reaction in conserved quantities. Therefore the capital equipment employed is included in both the bundle of inputs as well as the bundle of outputs. Each process can therefore be described by fixed input coefficients and fixed output coefficients just like a chemical reaction. Capital goods at different stages of 'wear and tear' can be treated as different goods. Furthermore it is assumed that consumption of goods takes place only through the processes of production, which include all necessities of life consumed by workers, and all income above necessities of life is reinvested in production. In an economy there are a fixed amount of processes and products representing all the technologically possible transformations, although not all processes need be active. Each process can be considered to be of unit time duration and longer processes can be broken down into a number of intermediate processes, with intermediate products if necessary.

The original von-Neumann model however is a static equilibrium model and analysis proceeds by considering relations which must hold at equilibrium. For example suppose the system has M processes labeled $i = 1, ..., M$ and N products labeled $j = 1, ..., N$. Suppose the input stochiometric ratios are given by a_{ij} for process i input of product j and b_{ij} for process i output of product j. $z(t)$ is an N dimensional vector representing the 'intensity' of processes i and $P(t)$ is a M dimensional vector describing prices of products j. Also suppose the present interest rate is given by $\beta(t)$. Several equilibrium relationships can be set up between these quantities.

(i) In equilibrium there can be no process which yields a return greater than the present interest rate for under perfect competition positive profits would attract competitors to use the same process so that prices of factors would rise. Therefore von-Neumann obtained,

$$BP(t+1) \leq \beta(t)AP(t) \tag{1}$$

where the output coefficient matrix is $B = (b_{ij})$ and the input coefficient matrix is $A = (a_{ij})$. This inequality says that the return of processing, i.e. the per unit of processing value of output divided by the per unit of processing cost of input must be less than or eaqual to the interest rate.

However if a process yields negative profits after payment of interest it will not be used, and its intensity is zero, therefore von-Neumann obtained,

$$z(t)BP(t+1) = \beta(t)z(t)AP(t). \tag{2}$$

(ii) Since each process is of unit time duration, the components of the vector $z(t-1)B$ give the amounts produced at time t, while those of the vector $z(t)A$ give the amounts of input used up in production at time t. It is impossible to consume more of a good in the production processes than is available, so von-Neumann gets,

$$z(t-1)B \geq z(t)A. \tag{3}$$

However in equilibrium those goods that are overproduced will be free goods and zero prices are charged for them. This implies,

$$z(t-1)BP(t) = z(t)AP(t). \tag{4}$$

Using these relationships von-Neumann is able to show a "state of balanced growth" where prices and rate of interest are constant over time and intensities of production grow or decay at constant geometric rates.

5.3. Dynamical Production Model

While the von-Neumann analysis is perfectly reasonable in equilibrium, in this paper we wish to consider disequilibrium. Indeed it is clear that real economies are usually far from equilibrium, with fluctuating prices, unemployment, cycles of recession and growth etc being characteristic. We hope to understand this ubiquitous behaviour.

One reason for disequilibrium in economic production is technological evolution. We do not consider technological evolution in this paper since our production coefficient matrices, like the VNM are fixed. However even without technological evolution one should not expect to find static equilibrium behaviour. For example the situation may be similar to the chemistry describing biological cells. Although the reaction matrix describing the internal cellular chemical reactions is fixed, the chemical concentrations needn't be fixed and at equilibrium, in fact the chemical concentrations are usually continuously varying, periodically or even chaotically[26]. Another well known example is provided by the populations of species in an ecosystem. Even though the matrices describing the predator prey relationships are fixed, the populations themselves are well known not to be at fixed point equilibrium, or balanced growth, but may indeed cycle or even show deterministic chaos. It is therefore natural that an economy, even with a fixed interaction matrix which ignores technological evolution, will show non-trivial dynamics.

In order to discuss disequilibrium we need to start from a bottom-up type of approach by considering the dynamics of a production process in its current environment. We may not impose top-down optimization equilibrium relationships to obtain solutions. For example in the above equilibrium

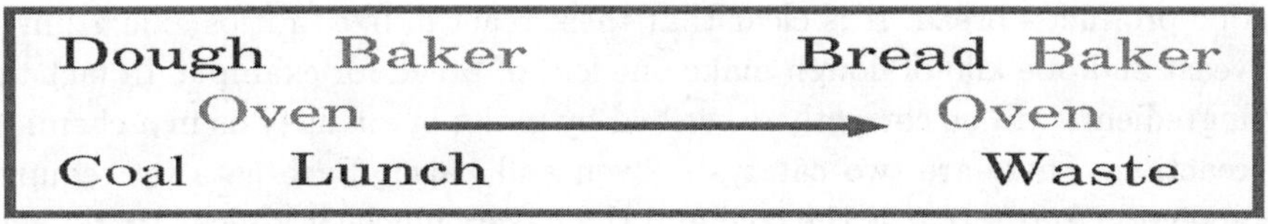

Fig. 5.2. Economic production processes are treated like chemical reactions where capital goods are catalysts. Consumption of goods necessary for the operation of the catalysts is included in process. The stochiometric ratios these products 'react' in are fixed and, here, the baker and oven are catalysts. The process may be further broken down into intermediate steps like a chemical reaction. For example the baker eats his lunch, becomes an 'activated' baker, puts the coal into the oven so the oven becomes activated, then puts the bread into the oven. Then the baker and oven decay back to the deactivated state accompanied by the production of bread and waste products.

analysis, rule (i) states that unprofitable processes will have zero intensity. In fact in disequilibrium processes may dynamically vary in profitability and intensity having profitable periods and loss making periods, as is the case in reality. Indeed rule (i) implies that entrepreneurs are perfectly rational and possess perfect foresight. In fact which processes are profitable at any given time may not be known by the entrepreneur since the process itself will affect its environment in a way that the entrepreneur may not, even in principle, be able to account for. By changing processing intensity, for example, the process will affect the supply and demand relationships and therefore the prices of the products it consumes or produces. A small change may be amplified by the complex non-linear feedback relationships between the different processes which interact and it will not be possible to treat some processes as profitable and others as unprofitable, as would be the case if their environments and the prices of their products were fixed and given. The processes will be in dynamical competition and symbiosis like species in an ecosystem.

Indeed in this model we will start from the bottom by modeling a production process itself in its dynamically changing environment and hope that such optimizing rules as (i) and (ii) will emerge as consequences of the dynamics. We try to make the simplest model possible to describe this situation. The system has two parts, production, where commodities are consumed and produced, and trading, where they are exchanged between processes.

We start with the production part. An example of an economic production process could, for example, is shown in Fig.5.2. In this process, which converts dough into bread, there are two ingredients, dough and yeast and one product - bread. It is clear that these react in fixed ratios, one gram of yeast and one kilo of dough make one loaf of bread for example. In fact the ingredients will be correctly described by moles of quantity as in a chemical reaction. There are two catalysts, oven and baker. Each catalyst requires an energy term, that is the coal and the bakers lunch. It is clear that where catalysts are concerned the relevant quantities are rates of operation. One baker can make all the bread in the world providing he is given long enough. Similarly the oven can bake it all. (Of course we are neglecting "wear and tear" in this first description.) So already the notion of time has entered the picture, and we see that it is impossible to consider production machines without considering time. In this case the baker and the oven both have maximum rates of production. The oven can bake 50 loaves an hour, and the baker load 80 loaves an hour, say. However maximum rates of production do not necessarily fix the production rates, this is given by the energy terms. In fact the baker can only load 80 loaves an hour if he is fed at least one bakers lunch a day. If he is not fed as much he can not work as much. Similarly the oven requires two sacks of coal an hour. Therefore if we start an hour period with 2 bakers, 5 bakers lunches, and plenty of dough, yeast, ovens and coal we can only make 2×80 loaves of bread in that hour. If we only have one sack of coal we can only make 25 loaves in the hour. If we have plenty of everything but only one oven we can only make 50 loaves in the hour. We see that in any given time period the amount of possible production is fixed as the *minimum* over the set of quantities at the start of the period. We can put the production machines into this picture by multiplying the quantity of production machines by the maximum rate at which they work by the length of the time period to give the production capacity in that time period. Putting these quantities together we get the quantity of production in one time period $\Delta t = 1$ as,

$$M_i(t) = Min(\frac{S_{i1}(t)}{a_{i1}}, \frac{S_{i2}(t)}{a_{i2}}, ..., \frac{S_{iN}(t)}{a_{iN}}) = Min(\frac{S_{ij}(t)}{a_{ij}}). \tag{5}$$

where $S_{ij}(t)$ denote the quantites of commodity j process i has at the start of the production period and a_{ij} are the input production coefficients. The $Min(x_i(t))$ denotes the minimum over the set $x_i(t)$.

Now we find that at the end of the production period process i has quantities $S_{ij}(t) + (b_{ij} - a_{ij})M_i(t)$ of product j where b_{ij} are the appropriate output production coefficients. For catalysts without any wear and tear $b_{ij} = a_{ij}$. The total supply of commodity j available in the whole economy after processing is given by $S_j(t) + \sum_i (b_{ij} - a_{ij})M_i(t)$, where

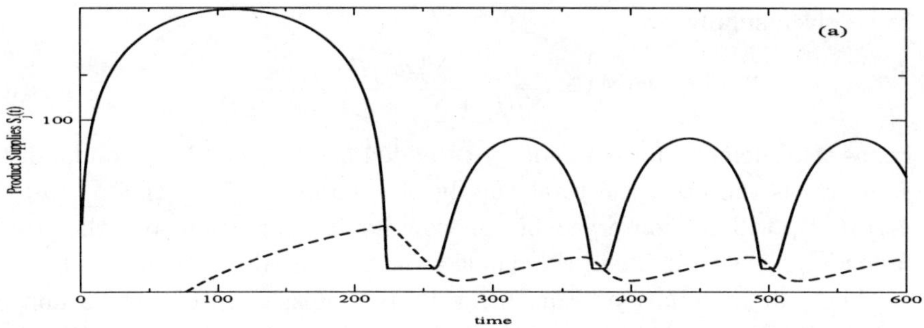

Fig. 5.3. (a) Single process supplies time series, $S_I(t)$ solid, $S_C(t)$ dashed. The parameters are $S_I^{ext} = 10$, $D_O^{ext} = 100$, $S_C^{ext} = 0.1$, $D_C^{ext} = 1.5$ and $D_I^{ext} = S_O^{ext} = 0$. $a_I = a_C = b_C = b_O = 1/2$.

$S_j(t) = \sum_i S_{ij}(t)$.

This describes the production part, now we need to consider how processes exchange products and allocate this available total supply between themselves. To facilitate this, in our model we also assume each process i has a value, $F_i(t)$, which we denote *funds*. This is the revenue of sales of its produced commodities in the preceeding timestep and therefore reflects the value (or *fitness*) of the process in the economy at that time. The funds $F_i(t+1)$ is then given by,

$$F_i(t+1) = \sum_j (S_{ij}(t) + (b_{ij} - a_{ij})M_i(t))p_j(t+1) \tag{6}$$

where $p_j(t+1)$ is the relevant price of product j. Note that this equation includes the quantities of product j produced, i.e. after processing.

Now in order to allocate the available supply we assume that each process uses these funds available from the previous period of processing, $F_i(t)$, to form *demands* for product j by dividing their funds according to the ratios, a_{ij}, that is according to the ratios needed for processing. Therefore process i forms demands $F_i(t)a_{ij}/\sum_j a_{ij}$, for product j. If we consider the a_{ij} normalized, $\sum_j a_{ij} = 1$, the demands simplify to $F_i(t)a_{ij}$. New supplies $S_{ij}(t+1)$ are then obtained as,

$$S_{ij}(t+1) = \frac{a_{ij}F_i(t)}{p_j(t+1)}. \tag{7}$$

In order for Eqs.6 and 7 to hold consistently we require the commodity j price $p_j(t+1)$ formed in the economy by this exchange of commodities to be given simply by,

$$p_j(t+1) = \frac{\sum_i a_{ij} F_i(t)}{S_j(t) + \sum_i (b_{ij} - a_{ij}) M_i(t)} \tag{8}$$

i.e. as total demand for commodity j divided by total supply of commodity j. If this is the case, the total supply of commodity j, $S_j(t) + \sum_i (b_{ij} - a_{ij}) M_i(t)$, will be conserved in the commodity allocation and the total funds $\sum_i F_i(t)$ available in the whole economy will also be conserved.

The pair of equations Eqns.6,7 with the subsitution of Eqns.5 and 8 define the temporal evolution of the simplest version of our model, and using them we calculate the development path of the economy.

We note that in considering the process i demand $F_i(t)a_{ij}$, for product j, the price of product j does not enter the formular. It is not difficult to incorporate knowledge of the last known price $p_j(t)$ in the demand forming equation, so that the process i demands for product j are given by, $F_i(t)a_{ij}p_j(t)/\sum_j a_{ij}p_j(t)$, rather than $F_i(t)a_{ij}$. In the following, since the feedback from the price is explicit in the demands formation process we call this the *'price feedback'* model, although it should also be remembered that simpler model also includes a more implicit price feedback through Eqn.8.

Indeed we note that we are free to consider any fund allocation $F_i(t)\sigma_{ij}(t)$, with the only requirement that $\sum_j \sigma_{ij}(t) = 1$, although we do not address these more general models further in this article.

5.4. Model Behaviour

Many interesting results have already been obtained, which we now summarize.

5.4.1. *Single Process in Fixed Environment*

In previous papers we have discussed the simplest possible case, that of a single process in a fixed environment[3,2]. Even the simplest one catalyst product, one input product, one output product, single process in a fixed environment is shown to oscillate with periods of inefficiency, where it has excess unused catalyst, and periods of full production. Such a system could represent a highly idealized single country where the catalyst may be the labour force and the inputs and outputs are imports and exports from abroad. The oscillation originates from the complex way the process itself *feeds back* onto the prices of the products it demands and supplies, and may be the basic reason for the ubiquitous national economic cycles with

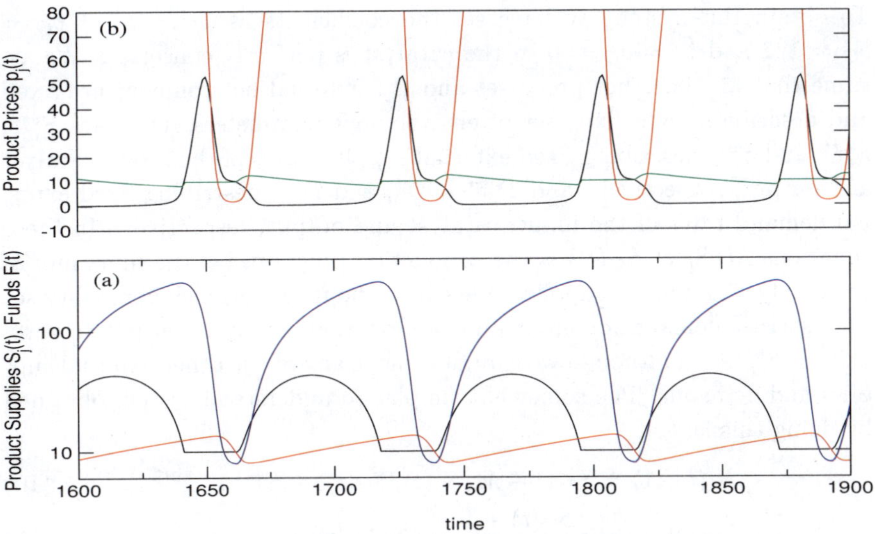

Fig. 5.4. (a) Single process in fixed environment supplies time series, input supply $S_I(t)$ with quadratic behaviour and catalyst supply $S_C(t)$ growing exponentially. Also shown is the process funds $F(t)$ which has largest variation. (b) Corresponding time series of prices, input price $p_I(t)$ showing sharp bubbles, catalyst price $p_C(t)$ (with cut off top), output price $p_O(t)$ (roughly constant).

periodic bouts of unemployment[3,2]. Here we briefly review that case and then discuss the modified case with explicit price feedback.

This simplest single process model has one input product I, one catalyst product C and one output product O. It has three variables, its supplies of input $S_I(t)$, its supplies of catalyst $S_C(t)$ and its funds $F(t)$, and is described by,

$$
\begin{aligned}
S_I(t+1) &= \frac{1/2F(t)}{p_I(t+1)} = 1/2F(t)\frac{S_I(t) - M(S_I(t), S_C(t)) + S_I^{ext}}{1/2F(t) + D_I^{ext}}, \qquad (9)\\
S_C(t+1) &= \frac{1/2F(t)}{p_C(t+1)} = 1/2F(t)\frac{S_C(t) + S_C^{ext}}{1/2F(t) + D_C^{ext}},\\
F(t+1) &= (S_I(t) - M(S_I(t), S_C(t)))p_I(t+1)\\
&\quad + S_C(t)p_C(t+1) + M(S_I(t), S_C(t))p_O(t+1)\\
&= \frac{S_I(t) - M(S_I(t), S_C(t))}{S_I(t) - M(S_I(t), S_C(t)) + S_I^{ext}}(1/2F(t) + D_I^{ext}),\\
&\quad + \frac{S_C(t)}{S_C(t) + S_C^{ext}}(1/2F(t) + D_C^{ext}) + \frac{M(S_I(t), S_C(t))}{M(S_I(t), S_C(t)) + S_O^{ext}} D_O^{ext}.
\end{aligned}
$$

To obtain this system we have set the coefficients as $a_I = a_C = b_C = b_O = 1/2$ and $b_I = a_O = 0$ so the catalyst is perfectly conserved. We assume that all the other processes and the external environment are fixed and describe it by a fixed set of environment parameters. They are S_I^{ext}, S_C^{ext}, and S_O^{ext} describing fixed external supplies rates of the input, catalyst and output, respectively, and D_I^{ext}, D_C^{ext}, and D_O^{ext} describing fixed external demand rates of the input, catalyst and output, respectively. In these equations $M(S_I(t), S_C(t))$ is the production function, i.e. the minimum of the catalyst and input supplies. We can without losing much generality set the external demand for input and the external supply of output to zero, $D_I^{ext} = S_O^{ext} = 0$, although we cannot remove any of the other external supplies and demands. The somewhat simpler to understand system obtained by doing this is,

$$S_I(t+1) = S_I(t) - M(S_I(t), S_C(t)) + S_I^{ext} \tag{10}$$

$$S_C(t+1) = \frac{S_C(t) + S_C^{ext}}{1/2F(t) + D_C^{ext}} 1/2F(t), \tag{11}$$

$$F(t+1) = \frac{S_I(t) - M(S_I(t), S_C(t))}{S_I(t) - M(S_I(t), S_C(t)) + S_I^{ext}} 1/2F(t) + \frac{S_C(t)}{S_C(t) + S_C^{ext}} (1/2F(t) + D_C^{ext}) + D_O^{ext}. \tag{12}$$

The oscillatory behaviour of this system is shown in Fig.5.3. Indeed at the root of this oscillatory behaviour is the interelationship of prices, process value, and processing rate, the minimum condition. In the process value equation, Eq.12, we see that some of the process value $F(t)$ is made up of the input product term $(S_I(t) - M(S_I(t), S_C(t)))p_I(t+1)$. This term is zero when the input supply $S_I(t)$ is the minimum $S_I(t) < S_C(t)$, since in this case all input is used up in processing. However when the catalyst supply $S_C(t)$ is the minimum this term reflects the value of unused excess input supply. This unused excess supply is unavoidable because the external rates of supply of input and catalyst, S_I^{ext} and S_C^{ext}, are not necessarily compatible with the process stochiometric ratios and in this simplest model the process simply divides its funds $F(t)$ into the stochiometric ratios (half and half here) in order to buy necessary catalyst and input material. As the process funds $F(t)$ increases, the process increases its catalyst supply $S_C(t)$ and the production rate $M(S_I(t), S_C(t)) = S_C(t)$ correspondingly increases. One might expect the contribution to process value $F(t)$ from the term $(S_I(t) - M(S_I(t), S_C(t)))p_I(t+1)$ to correspondingly decrease. However this is not the case. In fact the process value $F(t)$ is kept "artificially" high by the increasing *bubbling* price. The price of the input product is given by,

$$p_I(t+1) = \frac{1/2F(t)}{S_I(t) - M(S_I(t), S_C(t)) + S_I^{ext}}.$$

As $(S_I(t) - M(S_I(t), S_C(t)))$ decreases by processing, the price $p_I(t)$ will increase to balance the lower available supply and the process value $F(t)$ will remain high. In spite of its dwindling supplies of input $S_I(t)$ the factor in the input supply term in Eq.12, $\alpha(t) = (S_I(t) - M(S_I(t), S_C(t)))/(S_I(t) - M(S_I(t), S_C(t)) + S_I^{ext})$ will remain approximately 1 and the process will remain highly valued. When $S_I(t)$ becomes the minimum therefore there will be a sudden correction as $\alpha(t) = 0$. When the funds $F(t)$ collapses when $\alpha(t) = 0$ the collapse in catalyst $S_C(t)$ will follow and the rate of processing will become very low again and $S_I(t)$ will build up again, producing the cycle of boom and bust shown in Fig.5.3. We see that the economy is made up of a system of highly unstable interacting threshold terms.

In the discussion so far we considered the case where the demand ratios were fixed, i.e. the process allocated half of its funds $F(t)$ to buying input supply and half to buying catalyst supply. The obvious flaw in this set up is that in reality a process would consider the varying prices of the input and catalyst when allocating its available funds. If the catalyst had a much higher price than the input for example,in an effort to keep its supplies of catalyst and input in the correct stochiometric ratios the process would try to allocate more funds to the catalyst and less to the input. Could this perhaps remove the cycle of bubble and crash? The answer is no, in fact doing so makes the situation worse. The rationale now is that as the price of a commodity increases the process allocates more of its funds to demanding it, this increases the price even further. By incorporating the price into the demands we add another positive feedback loop onto the cycle.

Therefore in a very rough non-rigourous way we now briefly consider the modified system, which we call the *explicit price feedback* system,

$$S_I(t+1) = \frac{\sigma_I(t)F(t)}{p_I(t+1)}, \qquad S_C(t+1) = \frac{\sigma_C(t)F(t)}{p_C(t+1)}$$

$$F(t+1) = (S_I(t) - M(S_I(t), S_C(t)))p_I(t+1) + S_C(t)p_C(t+1) + M(S_I(t), S_C(t))p_O(t+1)$$

whose time series is described in Fig.5.4. Here the demand ratios for input and catalyst are $\sigma_I(t)$ and $\sigma_C(t)$, given by,

$$\sigma_I(t) = 1 - \sigma_C(t) = \frac{p_I(t)}{p_I(t) + p_C(t)} \tag{13}$$

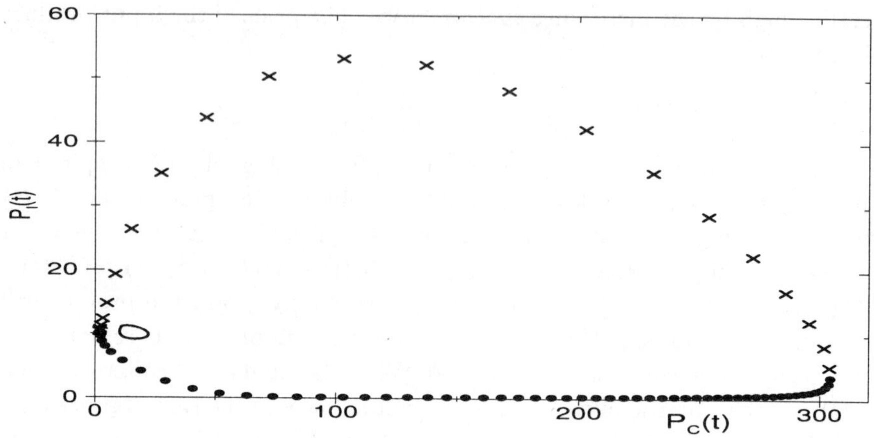

Fig. 5.5. The small cycle shows the variation of catalyst price $p_C(t)$ and input price $p_I(t)$ in the $p_C(t), p_I(t)$ plane in the case without explicit price feedback. The other time series is the case with explicit price feedback. The points are shown as circles when the input supply is in excess and crosses when the catalyst supply is in excess. As can be seen when a supply is in excess its price is near zero. This result expected from rational consideration only occurs with price feedback.

In the simplest oscillating process obtained again by setting $D_I^{ext} = S_O^{ext} = 0$ the prices are given by,

$$p_I(t+1) = \frac{\sigma_I(t)F(t)}{S_I(t) - M(S_I(t), S_C(t)) + S_I^{ext}} = \frac{p_I(t)}{p_I(t) + p_C(t)} a(t)$$

$$p_C(t+1) = \frac{\sigma_C(t)F(t) + D_C^{ext}}{S_C(t) + S_C^{ext}} = \frac{p_C(t)}{p_I(t) + p_C(t)} b(t) + c(t)$$

If we consider the background $a(t)$ and $b(t)$ slowly varying, the prices have two "fixed points". One fixed point is at $p_I^* = 0, p_C^* = b(t) + c(t)$ and the other is at $p_I^* \sim a(t), p_C^* \sim c(t)$ if $a(t) >> c(t)$. The stability of the fixed points roughly depends on the ratio $a(t)/b(t) > 1$, which is simply given by,

$$S_C(t) + S_C^{ext} > S_I(t) - M(S_I(t), S_C(t)) + S_I^{ext} \tag{14}$$

The RHS of Eq.14 is the supply of input product and the LHS is the supply of catalyst. In fact when the supply of a product is in excess its price is attracted to zero. Therefore the demand coefficient $\sigma(t)$ of the product in excess is approximately zero and all the process funds $F(t)$ is concentrated on demanding the product which is insufficient and the process does not demand the product in excess. While this may be what one would expect from rational expectation, it also makes bubbles and crashes stronger. Indeed if the input is excess we have $\sigma_I(t) \sim 0$, $p_I(t) \sim 0$ and $\sigma_C(t) \sim 1$ and we now find that the single process system is approximately given by,

$$\begin{aligned} S_I(t+1) &\sim S_I(t) - M(S_I(t), S_C(t)) + S_I^{ext} \\ S_C(t+1) &\sim (S_C(t) + S_C^{ext}) \frac{F(t)}{F(t) + D_C^{ext}} \\ F(t+1) &\sim (F(t) + D_C^{ext}) \frac{S_C(t)}{S_C(t) + S_C^{ext}} + D_O^{ext} \end{aligned}$$

Again this state describes the funds $F(t)$ and catalyst supply $S_C(t)$ growing slowly while the input supply $S_I(t)$ gradually decreases as the production rate $M(S_I(t), S_C(t)) = S_C(t)$ increases. Eventually the price fixed points stability, Eq.14 changes when the minimum switches so that $M(S_I(t), S_C(t) = S_I(t)$, and a fast attractor switch occurs to a new state where the input price $p_I(t)$ suddenly "bubbles" and the catalyst price collapses $p_C(t) \sim c(t)$. Then $\sigma_C(t) \sim 0, \sigma_I(t) \sim 1$ and we find,

$$\begin{aligned} S_I(t+1) &\sim S_I(t) - M(S_I(t), S_C(t)) + S_I^{ext} = S_I^{ext} \\ S_C(t+1) &\sim 0 \\ F(t+1) &\sim F(t) \frac{S_I(t) - M(S_I(t), S_C(t))}{S_I(t) - M(S_I(t), S_C(t)) + S_I^{ext}} + D_O^{ext} \sim D_O^{ext} \end{aligned}$$

As can be seen from these equations now the catalyst supply $S_C(t)$ quickly collapses due to the switch in demands for input and catalyst. This sudden collapse in $S_C(t)$ affects the price stability ratio $a(t)/b(t)$ Eq.14 and causes a sudden switch back to the zero input price phase. The effect is to produce a very fast bubble and then sudden crash in the input price. This rapid bubble and crash can be seen in Fig.5.4.

Although price oscillations occur in both cases, with and without explicit price feedback, the effect is much stronger in the explicit price feedback case. These results are confirmed in Fig.5.5 which plots the price variations in the $p_C(t), p_I(t)$ plane for both cases.

5.4.2. *Multiple Timescales*

Dynamics with simultaneous multiple timescales is typical for this model. Supply 'chains' have novel multiple timescale limit cycle attractors. The different timescales are controlled by the different distinct catalysts, as described in Fig.5.6. This dynamics is explained in more detail in Ref.[3]. Understanding this behaviour may be useful for the efficient control of factory

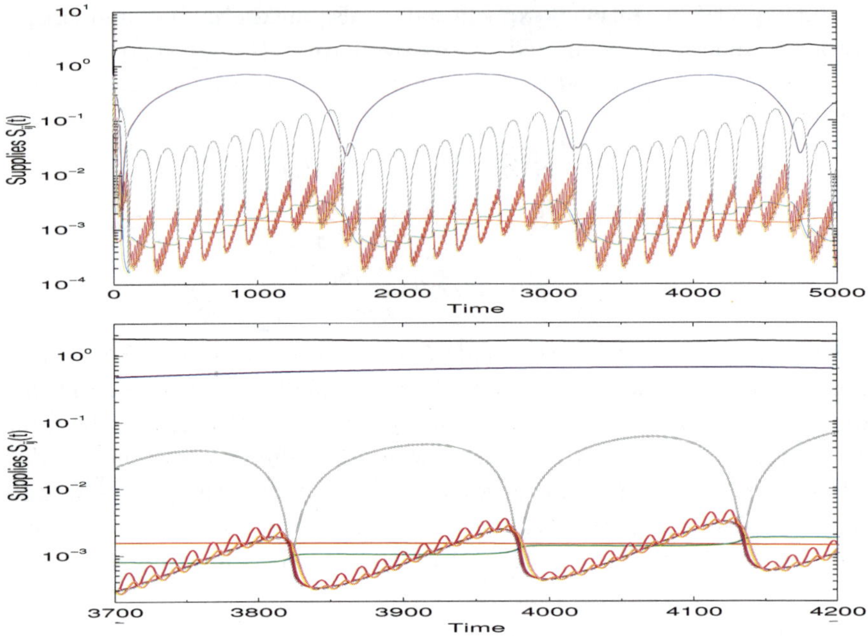

Fig. 5.6. Time series of supplies from a 4 process chain. Lower shows detail of upper. The chain is made of up 4 processes each with one input and one catalyst and one output. The output of each process provides the sole input for the next process in the chain. The external rates of supply and demand for all catalysts are fixed as are the external supply of the initial input product and external demand of the final output product. This shows time series of the 8 supplies, one input and one catalyst for each process. The different timescales are produced by the different catalysts in each process. Each catalyst has a different ratio of fixed external supply to fixed external demand, S_C^{ext}/D_C^{ext} giving each catalyst a different average 'external' fixed point price. The higher the catalyst price the slower the processing in the corresponding process and the slower the dynamical variation in the process. This is because higher catalyst prices mean lower catalyst supply rates and therefore lower processing rate, under the constraint that the average input supply rate is the same in each process and fixed by the external initial input supply rate.

supply chain dynamics. Similar problems have recently been studied by Helbing[18,19] and addressed to the design of semiconductor manufacturing plants[20]. Chains of supply processes with different price catalysts is also typical structure for a real economy, since most products are produced by a series of companies supplying parts to the next link in the chain. Therefore this this effect may be related to the appearance of complex business cycles on several timescales. This example illustrates that even in the case of a fixed environment, as is the case here, with fixed input supplies and demands for all products, the dynamics can be very complex and difficult to control.

A second simple example illustrates switching between timescales where the different timescales are controlled by different rate limiting catalysts. Shown in Fig.5.7 is the dynamical behaviour of the 3 process 4 product cyclic system, process 1: $A + B \rightarrow B + C$, process 2: $C + D \rightarrow D + B$, process 3: $B + D \rightarrow D + A$. The product D is not produced or consumed in this 3 reaction system, it just acts as a catalyst. Its external supply rates and demand rates are fixed so that $D_D^{ext} >> S_D^{ext}$ and therefore its average price is high. It is therefore in short supply and is the rate limiting minimum product in processes (2) and (3). This catalyst controls the slow timescale seen in Fig.5.7. The fast timescale is produced by the switching of the minimum in process (1), $M(S_{1A}(t), S_{1B}(t))$. Whenever the 'fixed points' of the average prices of A, $p_A(t)$ and B, $p_B(t)$ cross, B can become the rate controlling product in reaction (1) and the excess input supply $S_{1A}(t) - M(S_{1A}(t), S_{1B}(t)) = S_{1A}(t) - S_{1B}(t)$ becomes a new variable. This variable was zero when $M(S_{1A}(t), S_{1B}(t)) = S_{1A}(t)$ irrespective of the variation in $S_{1B}(t)$. The extra oscillation is produced by the new coupling of process (1) to process (2) and (3) through catalyst product B supply $S_{1B}(t)$. The frequency is very high because products A and B are produced in the same chain of processes and therefore $p_A(t) \sim p_B(t)$ so minimum switches in $M(S_{1A}(t), S_{1B}(t))$ occur very rapidly. This is explained in more detail in Ref.[3].

This dynamical behaviour may be relevant in biology. Indeed in biology, multiple timescales are also well known. Cells of course perform functions on many different timescales,[15,16,17]. Recently[14] up to 8 multiple timescales simultaneously occurring in the plasmodium of a single cell organism have been found. The dynamical origin of this may indeed be an adaptive coupling operating together with Michaelis-Menten type Leontief kinetics.

5.4.3. *Complex Dynamics*

The general dynamics of random networks of processes is extremely complex. Multiple timescale complex chaotic motion is typical. An example of such dynamics is shown in Fig.5.8. The time series shown is a section of the funds $F_i(t)$ time series for a random network of 15 processes. Each process consumes one input product and produces one output product. Each process also utilizes one catalyst. For each process these 3 products are chosen randomly from a set of 10 and we ensure that each consumed product

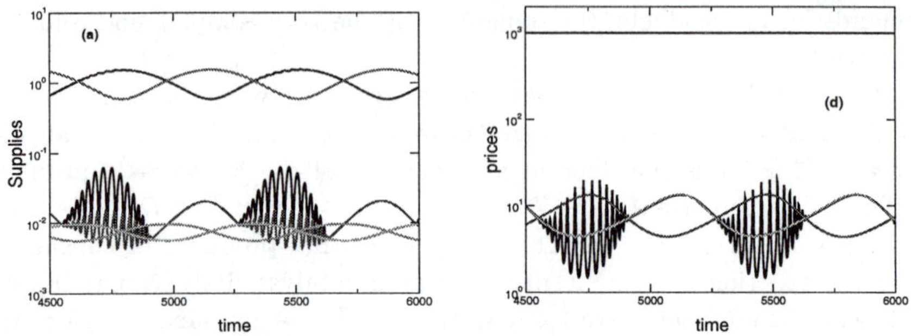

Fig. 5.7. Time series from three process cycle described in text There are three processes denoted 1, 2, 3 and four products A, B, C, D. Process 1 converts A into C via catalyst B. Process 2 converts C into B via catalyst D. Process 3 converts B into A via catalyst D. (a) The 6 individual process supplies. The high frequency oscillations occur in $S_{1A}(t)$ and $S_{1B}(t)$, which move around the same long term supply level as $S_{2D}(t)$ and $S_{3D}(t)$. The two supplies which move around much higher levels are $S_{2C}(t)$ and $S_{3B}(t)$. (b) Corresponding prices timeseries. $p_A(t)$ has high frequency oscillations and moves around the same level as , $p_B(t)$, and $p_C(t)$. This is because products A, B and C are produced and consumed by the same chain of processes. On the other hand product D has much higher price level, $p_D(t)$, controlled by the external environment.

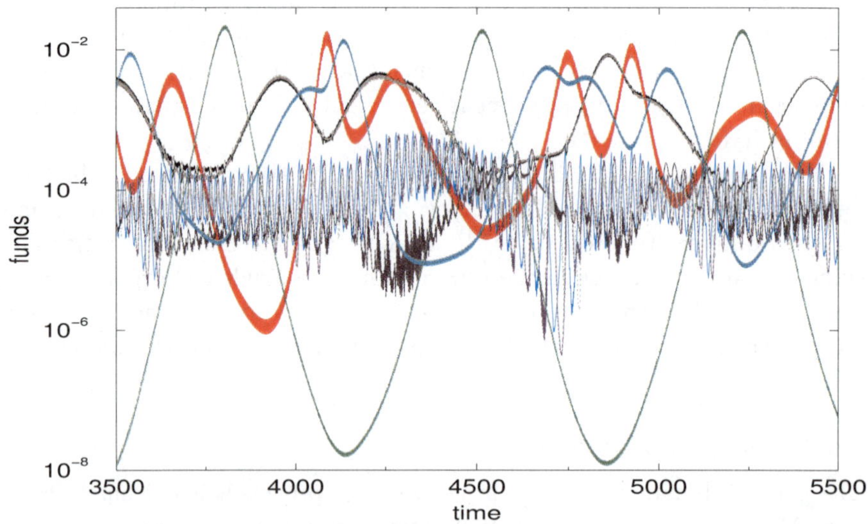

Fig. 5.8. Segment of time series from a 15 process 10 product random network where each process consumes one product, produces one product and employs a single catalyst product.

is produced somewhere else in the network and similarly each produced product is consumed somewhere. Even with this relatively small network of processes the dynamics is extremely complex and impossible to analyze completely.

In general this highly complex multiple timescale dynamics we call a minimum switching dynamics is created by the peculiarity of the minimum condition production function which divides the phase space into different regions obeying different systems of equations. In general when a process switches its minimum supply the catalytic pathways in the network change abruptly. For example in chemistry governed by the Law of Mass Action the pair of reactions $A + B \to B + C$, $C + D \to D + B$ is always autocatalytic. However in the minimum rate limited system this pair of reactions is only autocatalytic while $C < D$ (and $B < A$) so that C is rate controlling in the second reaction. Once C exceeds D the reaction rate is controlled by D which may not be catalysed by, or even depend on B, and whose supply may be controlled by an entirely different set of reactions elsewhere in the network. This effect 'limits' or 'binds' unstable dynamical motions, and allows different products to successively *switch control* of the rates of any particular sets of reactions. In a sense sets of reactions can have their own internal dynamical degrees of freedoms, such as C in the above pair, and the dynamical variation of other quantities such as D are entirely irrelevant to the dynamics of this pair unless $D < C$. The idea of the minimum switching being related to control can be seen more clearly by rewriting the equations for the product-process interaction as follows,

$$F_i^* \leftarrow 1/N(F_O + \sum_i F_i), \qquad S_i^* \leftarrow S_i - M(S_j)$$

$$F_i \leftarrow \frac{(S_i - M(S_j))(F_i^* + D_i^{ext})}{S_i - M(S_j) + S_i^{ext}}, \qquad S_i \leftarrow F_i^* \frac{S_i - M(S_j) + S_i^{ext}}{F_i^* + D_i^{ext}}$$

$$F_O^* \leftarrow 0, \qquad S_O^* \leftarrow M(S_j)$$

$$F_O \leftarrow \frac{M(S_j) D_O^{ext}}{M(S_j) + S_O^{ext}}, \qquad S_O \leftarrow 0$$

These mappings describe a process with several input products labeled $i = 1, .., N$ and one output product labelled O. Variables with superscript $*$ denote *outputs from* the process, while variables without the superscript denote the *inputs to* the process. Firstly the S_O^* mapping shows explicitly that the dynamical variation of the process output is *controlled* by only one input - the minimum input dynamics. Secondly the F_i^* and F_i mappings show explicitly that while the process sends funds to all its inputs equally,

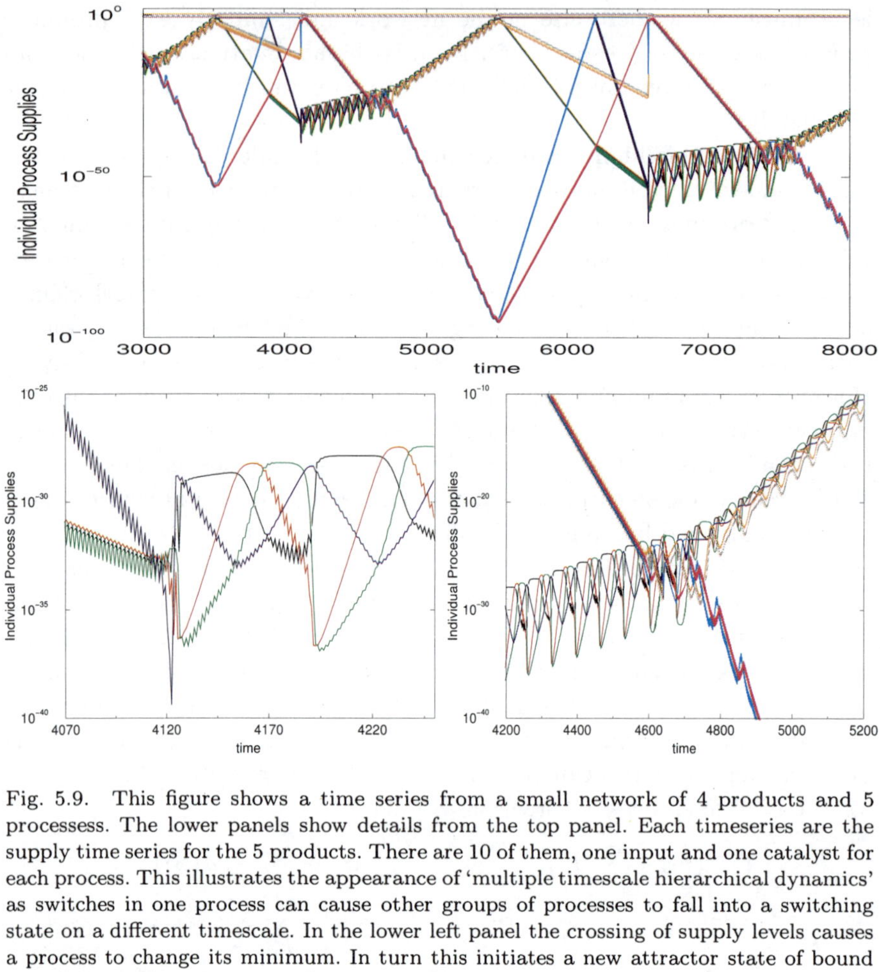

Fig. 5.9. This figure shows a time series from a small network of 4 products and 5 processess. The lower panels show details from the top panel. Each timeseries are the supply time series for the 5 products. There are 10 of them, one input and one catalyst for each process. This illustrates the appearance of 'multiple timescale hierarchical dynamics' as switches in one process can cause other groups of processes to fall into a switching state on a different timescale. In the lower left panel the crossing of supply levels causes a process to change its minimum. In turn this initiates a new attractor state of bound repeated switching in a pair of processes. As can be seen the switching itself binds the 4 levels together into a periodic oscillation state with its own characteristic frequency. However this oscillation is modulated by slower change in other parts of the network and as can be seen two different growing and decaying bound oscillating states eventually 'collide'. The collision shown in the lower right hand panel causes the initiation of a new attractor state where now 3 processes all switch their minimums in a different frequency bound state.

F_i^*, it receives funds back, F_i, in proportion to the excess unused input $S_i - M(S_j)$. This is the smallest (zero in fact) for the minimum input product. This means the *net* flow of funds F is always towards the minimum input supply product, that is the input product controlling the output variation.

Indeed because the excess supply $S_i - M(S_j)$ is zero for the minimum product it can be said that the process is *forced* by the minimum product. This is because the process sends exactly zero supply to the minimum product, S_i^*, and receives exactly zero funds, F_i from the minimum product. This is not true for non-minimum products and catalysts where there is a continuous flow of supply and funds in both directions between product and process. Indeed for the non-minimum products this continuous exchange produces an oscillation, such as seen above for the process with the single input, single output and single catalyst. The process is seen to *channel* funds towards the minimum input product, which is the one in control of the output variation. Although we have illustrated this point here for the system without explicit price feedback, the idea is the same (and much stronger) when explicit price feedback is included.

This observation that a minimum condition could be utilised in control could be relevant in biological systems. In particular it may be relevant for internal cellular control, implemented by Michealis-Menten threshold kinetics and also for neural switching dynamics[40].

An example of the complex dynamics this system can exhibit is shown in Fig.5.9. This is a small system composed of 5 processes and 4 products. In this example the dynamics includes a heteroclinic cycle which produces very small values for the variables. Therefore we do not expect this example to be relevant for physical or biological systems, it is simply intended as an illustration of the complexity that can be produced even in small systems. Indeed in larger systems of 10 to 15 processes and products (see Fig.5.8) we can produce dynamics of similar complexity but without heteroclinic cycles. The detailed dynamical origin of these dynamics will be addressed in future publications.

This example shown in Fig.5.9 illustrates how the minimum switching condition can 'bind' otherwise unstable dynamical variations into complex cycles. As can be seen the dynamical cycles occuring in sets of processes can be hierarchically controlled by minimum switches in other processes' which can turn the cycles on and off. This is because switching changes the catalytic feedback pathways in the network as described above. This allows different sets of processes to have their own 'internal' variables, temporarily decoupled from the rest of the network, and for different sets of processes to control each other, while the minimum binding limits unstable dynamical states. This kind of dynamics can be the basis for a kind of self organized information processing which could occur in biological systems. Indeed it has been pointed out that information processing necessarily requires the

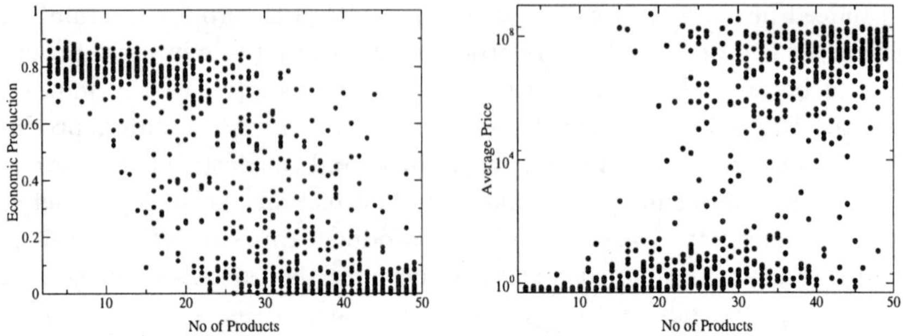

Fig. 5.10. (a) Variation of total economic production in the network as number of products is varied for fixed number of processes. This quantity is defined as $M = \sum_j M_j$ where the sum is over the production M_j in each process j. (b) Variation of average product price in the network as number of products is varied for fixed number of processes. In both figures each point is a long time average from one randomly chosen network of 15 processes and 10 products with random initial conditions, and a long transient was discarded.

cross-modulation of dynamical processes simultaneously occurring on different timescales, longer timescales acting as 'memory stores' modulating shorter timescales[35,36,37].

5.4.4. *Network Structure*

This network is adaptive as explained above. The adaptation is mediated by the flow of funds F through the processes throughout the network from external environmental demands to external environmental supplies. Although which processes are technologically possible is fixed in this model by the fixed input and output stochiometric ratio matrices as described above, so that we do not consider technological evolution, this does not mean that the network structure is fixed. In fact, besides the modulation of the 'size' of a production process by its fitness F the network is also 'prunable'. This is because processes also have a 'bankrupt' fixed point at the origin, $F^b = S_I^b = S_C^b = 0$ for example, in the case of the simplest single process described above. In the case of the simplest single process in a fixed environment this is a stable attracting fixed point attracting all initial conditions when $D_I^{ext}/S_I^{ext} > D_O^{ext}/S_O^{ext}$, irrespective of the external rates of supply and demand for the catalyst. This condition is what one might expect since it means that the external fixed point input price is higher than the external fixed point output price (these are the prices of the input and output products if the process didn't exist) and in this case the process will not be profitable. This means that in any network some processes may be completely inactive at any time (although they may reactivate later if the environment changes.)

Indeed we also find that some complete network structures are unstable and collapse. This collapse is accompanied by unstable (divergent) prices, low production and high inefficient unused catalyst, i.e. inflation and unemployment[1]. See Fig.5.10. Indeed we find a phase transition where networks are stable when there are many competing processes, i.e. capitalist, but complete collapse occurs in 'monopolistic' networks. Similar results have recently been confirmed by Martino et al[4] in a purely statistical treatment of production networks. The different dynamics associated with the different phases is currently under investigation. We find most complex time series occur in the transition region while in the capitalist phase, dynamics is simple, typically fixed point. This could relevant to the understanding and prediction of economic collapse and the failure of communism!

We have also studied the emergence of chains of processes. We find that as large networks adapt to their environment by some processes going bankrupt, a network of interlocking 'assembly lines', changing input into output and producing their own catalysts in the most efficient way, naturally emerges. These are the economic equivalent of catalytic sets. Catalysts in cells are known to operate in cellular assemblies in a similar way to factory assembly lines, and our emergent result, the result of an adaptive efficiency maximizing process may be a basic reason for this. The structure of yeast protein networks, showing peripheral 'hubs' for example, is recently under intense scrutiny and our model may be relevant to the explanation such structures from a functional viewpoint[34].

This network can also be useful for analysis of product marketing in complex economies and in this respect we are studying catalyst specialization and substitute death. When the technological network structure is such that a given input product may be converted into a given output product by several pathways with different catalysts, so that these represent substitute products for example, we sometimes find that only one catalyst is selected (the cheapest one - as mutually determined by the rest of the network structure of course) while the other pathways go bankrupt. In other circumstances however coexistence of substitute products is possible. Analysis of product coexistence may have use in real company marketing concerns and also to the well known Sony Betamax problem, addressed by W. Brian Authur, where the final winner in a competition may also be dependent on chance fluctuations in initial conditions.

This study is ongoing and may also may be relevant to understanding cellular biology where catalysts are well known to be specialized to certain substrates.

5.5. Discussion

There are many analogies between economic production networks and the internal organization of a living cell.

Indeed the last 20 years has seen a revolution in our understanding of cellular biology. The simplistic idea of a cell as a homogenous well-stirred container for chemical species randomly diffusing, colliding and reversibly reacting according to the Law of Mass Action[23], while energy rapidly dissipates throughout the entire volume, has long gone. In a recent article: 'The cell as a collection of protein machines: preparing the next generation of molecular biologists'[8], B.Alberts writes 'Instead of a cell dominated by randomly colliding individual protein molecules, we now know that nearly every major process in a cell is carried out by assemblies of 10 or more protein molecules. A cell can be viewed as a *factory* that contains an elaborate network of interlocking *assembly lines*, each of which is composed of a set of large protein *machines*.' The analogy between economic production networks and cellular biology runs very deep, and must be *driven by basic physical principles* underlying the growth of complexity and structure in both these systems. The basic physics may be embodied in the adaptive networks of production processes explained above.

For example proteins function very much like machines in factories and contain highly coordinated moving parts[9,10]. Indeed in enzyme catalysis the binding of a substrate molecule causes the enzyme-substrate complex to relax slowly to a different conformational state which facilitates a reaction. After the reaction the product leaves the enzyme and the enzyme returns slowly to its original conformational state.

Each protein is highly specialized to its particular task and usually only binds to one type of substrate[8,10]. This high specialization is an evolved characteristic of living cells in the same way as specialization appears in economy[38,39] through division of labour in human societies and the evolution of more and more complex and specialized machines. In fact in interesting new results, as explained above, our economic model shows *emergent* catalyst-substrate specialization due to process death. This result requires further investigation for applications to understanding protein specialization in cell biology.

As explained above the rate of an economic processes is given by the minimum of its input supplies and catalysts, not by the Law of Mass Action, since in economics there is no notion of randomness. In enzyme dynamics too, the law of mass action is not applicable. In fact, in cells the characteristic time for an enzyme-substrate conformational change and for the product liberation to occur is much slower than diffusion driven characteristic transport times of regulatory and substrate molecules[7,11,12]. Therefore processes inside individual reacting molecules are much slower than diffusion kinetics and the Michaelis-Menten reaction rate is much more appropriate. Furthermore the Michaelis-Menten form is of threshold nature and very similar in form to the minimum condition explained above. While the minimum condition is appropriate for economics, we also hope in this future research to generalise the network dynamics according the Michaelis-Menten form. We also note, as described above, that the threshold condition itself may be important in switching of control of multiple timescale temporal sequencing of cellular reaction processes between different catalytic sets leading even to a kind of computation.

Furthermore in our economic model, product exchange is assumed to happen instantaneously and globally through central markets. Since diffusion times in the cell, and therefore mixing times, are much faster than the protein conformational changes, the cell production processes can be considered thoroughly mixed or globally coupled[7,11,12]. Therefore for the cell too a globally coupled network is appropriate.

Protein machines do not usually act alone, but in protein assemblies. Within each protein assembly, intermolecular collisions are not only restricted to a small set of possibilities, but reaction C depends on reaction B which depends on reaction A like a factory production assembly line[8,10]. Like Toyota just in time production line-management products are made when they are required and then used up. In our economic model, assembly lines appear in a self-organized way by bankruptcy of competing processes using the same products or producing the same products[1].

It is well know that energy does not dissipate throughout the cell but remains localized in non-equilibrium protein conformations or stored in ATP for example. Indeed this is necessary since protein machines perform mechanical work and require an energy source. This energy may be moved around and released to energy accepting reactions by the energy donating mechanism of ATP hydrolysis when it is needed by particular production lines. This direct energy coupling mechanism ensures coupling of energy-accepting and energy-donating processes without equilibriating with the external degrees of freedom of the surrounding medium. In the economy *funds* or money may represent stored energy in a very similar way, and transfer of money between buyer and seller proceeds in exactly the same way, through direct coupling without dissipation.

As explained above, in economics, adaptation is brought about through the transfer of funds, in a cell adaptation may therefore even be carried out by the transfer of energy. Indeed such highly organized activity as the cellular assembly line described above must emerge as a natural consequence of interactions between various elements[7]. In fact this highly organized activity is brought about by ordered conformational changes in the proteins driven by ATP hydrolysis (or by other sources of energy, such as an ion gradient)[8]. This picture is intriguingly similar to the economic production chain explained above where demand in funds is fed back up the chain from external demand in funds to external supply of substrate.

Economic reactions as defined by our economic production network are irreversible. One cannot obtain dough from bread! In the cell too because the conformational changes are driven they dissipate free energy and generally proceed in only one direction[8,10], and irreversibility is characteristic.

Besides these analogies to cellular biology, this network model may also have relevance to other biological systems. Two obvious examples of candidate biological adaptive production process networks are the immune system, which is known to evolve or adapt, and the brain[40].

With regard to the brain, the environmental input supply may represent flow of sensory input 'signal' to the brain. The output supply S from the network to the environment would represent motor response signal. The environment would respond to this output motor signal with 'reinforcment reward' flow F (the external demands). Processes would represent interconnected neurons or neuronal groups. The minimum condition would direct the net flow of reinforcement through the neuronal groups towards the rate controlling minimum input, as described above. This could provide a self-organized means for the control of complex real time information processing as described above. It would also be a self-organized means for the reinforcement flow coming from the external environment to *target* the appropriate area of the network, since the area being targetted would automatically be the rate limiting minimum catalyst, or catalytic set, in the network which is limiting the flow of input sensory signal to output motor signal, and is therefore 'in charge of' the output signal variation. Indeed multiple timescale oscillations are well known in the brain although their functional significance is not well understood. They have been related to the idea of neuronal units 'gating' each other[28,27,33,30,31,29]. This model could provide an understanding for these oscillations as being caused by rate limiting 'decision making' neuronal catalytic sets. This idea is described in more detail in Ref.[40].

Acknowledgments

This work was supported by The Japan Society for the Promotion of Science. Adam Ponzi would like to thank the organizers of the 'Networks of Interacting Machines: Production Organization in Complex Industrial Systems and Biological Cells' conference for inviting this contribution and the Tschira Foundation for support. Adam Ponzi would also like to thank K.Kaneko and A.Yasutomi for much valuable input.

References

1. Ponzi, A., Yasutomi A., Kaneko K., (2003), *A Non-Linear model of Economic Production Processes,* Physica A, 324, 372-379.
2. Ponzi, A., Yasutomi A., Kaneko K., (2003), *Multiple Timescale Dynamics in Economic Production Networks,* nlin.AO/0309002.
3. Ponzi, A., Yasutomi A., Kaneko K., (2003), *Complex Dynamical Behaviour in Economic Production Networks* To appear in Journal of Economic Behaviour and Organization.
4. A. De Martino, M.Marsili, R.Mulet, *Typical Properties of Large Random Economies with Linear Activities*, cond-mat/0309533.
5. J.von Neumann, *Review of Economic Studies* **31** 1-9 (1945).
6. M.Morishima, *Theory of Economic Growth*, (Oxford University Press 1970).
7. A.S.Mikhailov, V.Calenbuhr, *From Cells to Societies, Models of Complex Coherent Action*, (Springer 2002).
8. B.Alberts, Cell, **92**, 291 (1998).
9. C.W.F.McClare, J.Theor.Biol. **30**, 1, (1971).
10. L.A.Blumenfeld, A.N.Tikhonov, *Biophysical Thermodynamics of Intracellular Processes: Molecular Machines of the Living Cell*, (Springer, Berlin 1994).
11. A.S.Mikhailov, B.Hess, J.Phys.Chem. **100**, 19059, (1996).
12. P.Stange, A.S.Mikhailov, B.Hess, J.Phys.Chem.B, **103**, 6111, (1999). J.Phys.Chem.B, **104**, 1844, (2000).
13. M.Schienbein, H.Gruler, Phys.Rev.E **56**, 7116, (1997).
14. T.Ueda (preprint).
15. A.Goldbeter, *Biochemical Oscillations and Cellular Rythms*, (Cambridge University Press, 1996).
16. J.D.Murray, *Mathematical Biology*, (Springer-Verlag, 1993).
17. A.T.Winfree, *The Geometry of Biological Time*, (Springer-Verlag, 1980).
18. D.Helbing, *Modelling Supply networks and business cycles as unstable transport phenomena*, New Jounal of Physics **5**, 90, (2003).

19. T.Nagatani, D.Helbing, *Stability analysis and stabilization strategies for linear supply chains*, Physica A 335, 644-660.
20. D.Helbing, *Modelling and Optimization of Production Processes*, 85-105 in G.Radons and R.Neugebauer (eds) Nonlinear Dynamics of Production Systems (Wiley-VCH, Weinheim.)
21. K.Kaneko & T.Yomo, *Bull.Math.Bio*, **59** 139-196 (1997).
22. C.Furusawa & K.Kaneko, *J.Theor.Bio*, **209** 395-416 (2001).
23. P.W.Atkins, *Physical Chemistry*, (Oxford University Press 1994).
24. Neumann, J.von, 1945, *A Model of General Economic Equilibrium*, Review of Economic Studies, 31, 1-9.
25. Morishima M., 1970, *Theory of Economic Growth,* Oxford University Press.
26. Frusawa C., Kaneko K., 2001, *Theory of Robustness of Irreversible Differentiation in a Stem Cell System: Chaos Hypothesis*, Journal of Theoretical Biology, **209** 395-416.
27. M.G.Murer, K.Y.Tseng, F.Kasanetz, M.Belluscio, L.A.Riquelme, Brain Oscillations, Medium Spiny Neurons and Dopamine, Cellular and Molecular Neurobiology, 22, 5, 611, (2002).
28. J.D.Berke, M.Okatan, J.Skurski, H.B.Eichenbaum, Oscillatory Entrainment of Striatal Neurons in Freely Moving Rats, Neuron Vol 43, 883-896 (2004).
29. S.Keri, Units and rhythms of the brain are revealed., Neuroreport Vol 15, 1231, (2004).
30. Mizuhara H, Wang L-Q, Kobayashi K, Yamaguchi Y, Emergence of a long-range coherent network associated with frontal midline theta during a mental task in human: a study of simultaneous EEG and fMRI. Neuroreport 2004; 15:1233-1238.)
31. N.Sato, Y.Yamaguchi, Memory Encoding by Theta Phase Precession in the Hippocampal Network, Neural Computation, 15, 2379 (2003).
32. J.E.Lisman, M.A.Idiart, Storage of 7 ± 2 short term memories in oscillatory subcycles, Science, 267, 1512, (1995).
33. Y.Goto and P.O'Donnell. Network Sychrony in the Nucleus Accumbens in Vivo. Journal of neuroscience, June 2001, 21(12) 4498.
34. S.Maslov, K.Sneppen, Topology of Molecular Networks, Science, 296, 910-913 (2002).
35. S.Wolfram, Universality and Complexity in Cellular Automata. Physica D, 10:1-35, (1984).
36. N.H.Packard, Adaptation toward the Edge of Chaos. In J.A.Kelso, A.J.Mandell, M.F.Shlesinger, editors, Dynamic Patterns in Complex Systems, 293-301, World Scientific, Singapore, 1988.
37. C.G.Langton, Computation at the Edge of Chaos: Phase Transitions and Emergent Computation. Physica D, 42:12-37, 1990.
38. Adam Smith, The Wealth of Nations.
39. A.Ponzi, A Model of Division of Labour, Proceedings of Santa Fe Institute CSSS02, (2002).
40. A.Ponzi, The Self-Organization of Multiple Timescale Dynamics in a Neural Reinforcement Flow Network, Riken BSI internal paper, February 2005.

CHAPTER 6

UNIVERSAL STATISTICS OF CELLS WITH RECURSIVE PRODUCTION

Kunihiko Kaneko[1,2], Chikara Furusawa[3,2]

(1) Department of Pure and Applied Sciences, Univ. of Tokyo, 3-8-1 Komaba, Meguro-ku, Tokyo 153-8902, Japan

E-mail: kaneko@complex.c.u-tokyo.ac.jp

(2) ERATO Complex Systems Biology Project, JST, 3-8-1 Komaba, Meguro-ku, Tokyo 153-8902, Japan

(3) Department of Bioinformatics Engineering, Graduate School of Information Science and Technology, Osaka University, 2-1 Yamadaoka, Suita, Osaka 565-0871, Japan

E-mail: furusawa@ist.osaka-u.ac.jp

Statistical behavior of chemical reaction dynamics in a reproducing cell is studied. Zipf law in rank-abundance relationship in chemical compositions and log-normal distribution of abundances over many cells are universally found for a cell that exhibits recursive production, and theoretically explained. Regulation of such large abundance fluctuations, as well as relevance of the statistics to biological and social systems is briefly discussed.

6.1. Question to be Addressed

A cell consists of several replicating molecules that mutually help the synthesis and keep some synchronization for replication. At least a membrane that partly separates a cell from the outside has to be synthesized, keeping some degree of synchronization with the replication of other internal chemicals. Both catalysts and resource chemicals exist in some balance to maintain recursive production. How is such efficient recursive production possible while keeping diversity of chemicals? Is there some universal statistics in abundances of chemicals and in the network for a cell with steady recursive growth?

In a cell, a variety of chemicals form a complex reaction network to synthesize themselves. Then how such cell with a huge number of components and complex reaction network can sustain reproduction, keeping similar chemical compositions?

On the other hand, the total number of molecules in a cell is limited. If there is a huge number of chemical species that catalyze each other, the number of some molecules species may go to zero. Then molecules that are catalyzed by them no longer are synthesized. Then, other molecules that are catalyzed by them cannot be synthesized. In this manner, the chemical compositions may vary drastically, and the cell may lose reproduction activity.

Of course, the cell state is not constant, and a cell may not divide for ever. Still, a cell state is sustained to some degree to keep producing similar offspring cells. We call such condition for reproduction of cell as 'recursive production' or 'recursiveness'. The question we address here is if there are some conditions on distribution of chemicals or structure of reaction network.

The number of each molecule in a cell changes in times through reaction, and the number, on the average is increased for the cell replication. Positive feedback process underlying the replication process may lead to large fluctuations in the molecule numbers. With such large fluctuations and complexity in the reaction network, how is recursive production of cells sustained?[a]

6.2. Logic

Here, we study statistical characteristics that a cell with recursive growth has to satisfy. In a cell, there is a huge number of chemicals that catalyze each other and form a complex network. Through membrane, some chemicals flow in, which are successively transformed to other chemicals through this catalytic reaction network. For a cell to grow recursively, a set of chemicals has to be synthesized for the next generation. Each chemical, for its synthesis, requires some other chemicals as a catalyst. Then, generally speaking, there should exist some mutual relationship among molecules to catalyze each other.

---

[a]This type of problem has been discussed in relationship with the origin of life[1,2,3]. Here we discuss this problem in connection with a universal feature of a cell with recursive production, in general, which should hold also for the present cell.

While the number of molecule species is huge in a cell, the number of each molecule species is not necessarily large. Then fluctuations in molecule number are inevitable, since the reaction occurs through stochastic collision. The number of some molecules may go to zero, which sometimes may be dangerous, since such molecules may be essential as a catalyst to the synthesis of some molecules. Then, it is not trivial how recursive production of molecules in a cell is sustained.

Here it will be too much demanding to request that all the molecules keep their number through recursive production of a cell. The compositions of chemicals can differ by cells through divisions. Such loose reproduction should be all right as a beginning of cell. Still, with this loose reproduction, the catalytic activities should be sustained to keep reproduction of cells. In this sense we need to understand how an ensemble of molecules keep catalytic activity and loose reproduction in the amidst of large fluctuations in the chemical compositions.

The biochemical reactions for metabolism, synthesis of membrane and nucleic acid keep some synchrony. In a cell, some transported nutrients are successively transformed to some other chemicals that include catalysts for other reactions. If the transportation of nutrients is higher, that would be helpful for the growth. But if nutrients are too much there will be no room for catalysts from it, and the reaction no longer works. Various chemicals should exist in order for catalytic reactions to work efficiently. On the other hand, if all chemicals exist in the same order, first, the probability of each molecule to meet its catalysts will be lower, and the effective transformation of nutrients is not possible. Furthermore, in this equally distributed number of chemicals, since there are many molecule species that are low in concentration, the reaction events progress just randomly, and the chemical compositions will differ much by generations.

Thus, in order for effective transformation of nutrients, existence of some structure in abundances in chemicals should be favorable. Indeed, in catalytic reactions, there are successive structure, as catalysts for the reactions to transform nutrients, then catalysts for such catalysts for nutrients, and then catalysts for catalysts for catalysts for nutrients, and this cascade continues. It would then be expected that these levels of catalysts do exist in different levels of abundances. On the other hand, if such unbiased distribution in abundances of chemicals exist, the reaction probability is not homogeneous for each reaction. This will lead to decrease the random change of concentrations, as compared with the case of almost equal distribution in numbers.

Hence, for an effective use of resource chemicals, some hierarchical structure in catalysts is favored, with regards to the abundances. Indeed, by taking a specific cell model, we will explicitly show existence of such cascade structure. When a hierarchical structure exists, it is easily expected that some power law exists in the abundances. The distribution $\rho(x)$ that the abundances of chemicals is between x and $x + dx$ is then given by a power law. Indeed, as will be shown, from several model simulations, we find universal power law statistics for a cell that grows efficiently and recursively[4].

At this stage, however, we need to seriously consider the fluctuation of the number of molecules[5]. The chemical abundance of each molecule is under some fluctuation, since the collision process of molecules is basically stochastic. As long as the total number of molecules in a cell is not very huge, fluctuation in each molecule concentration are inevitable. Of course, negative feedback process to stabilize the concentration is one possible solution, to reduce the fluctuations. For reproduction of a cell, however, molecules have to be synthesized which implies some positive feedback process to amplify the number of each molecule species. As for growth, it is better to strengthen this amplification rate, which, however, may amplify the fluctuations also. We need to study some general features of the fluctuations inherent in such positive feedback system. As a very simple illustration, let us consider a process that a molecule x_m is replicated with the aid of other catalytic molecules.

Then, the growth of the number $N(m)$ of the molecule species x_m is given by $dN(m)/dt = AN(m)$. Here A involves the rate of several reaction processes to synthesize the molecule x_m. Such synthetic reaction process depends on the number of the molecules involved in the catalytic process. Recall, however, that all chemical reaction processes are inevitably accompanied with fluctuations arising from stochastic collision of chemicals. Thus, although the reactions to synthesize a specific chemical and convert it to other chemicals is balanced in a steady state, the fluctuation terms remain. Accordingly, the above rate A has fluctuations $\eta(t)$ around its temporal average $\overline{a}$. Then the above rate equation is written as $dN(m)/dt = N(m)(\overline{a} + \eta(t))$, and it follows that

$$dlogN(m)/dt = \overline{a} + \eta(t). \tag{1}$$

In other words, the logarithm of chemical abundances shows Brownian motion around its mean. The logarithm of chemical abundances is expected to obey normal (Gaussian) distribution, if $\eta(t)$ is approximated by a Gaussian noise, Accordingly, the logarithm of the number molecules is suggested to obey normal (Gaussian) distribution, as

$$P(N) \propto exp(-\frac{(logN - logN_0)^2}{2\sigma}). \tag{2}$$

Such distribution is known as log-normal distribution. In contrast to the Gaussian distribution, the log-normal distribution has a longer tail for more abundances, if plotted in the original scale without taking logarithm.

Of course, the present argument is too simple, since the reaction process is not necessarily directly autocatalytic, as assumed in the above argument. Replication of a molecule occurs only through several steps of reactions. Still, in each term of reaction process, there can appear a multiplicative stochastic process with $\eta(t)x_m$, for the change of x_m, in general. Hence, the log-normal distribution, rather than Gaussian distribution, may be common in a cell that reproduces itself recursively[7,6]. This argument on this fluctuation is rather primitive, and we need to check if it really works in a model and experiments. We will discuss this problem later.

As mentioned, the log-normal distribution has a very large tail to the abundant side. This large fluctuation in number, in some sense is quite far from the controlled behavior of a cell. Such fluctuations may destroy the recursive growth of a cell. Is there some control mechanism to suppress some fluctuations there? Is gene a controller for such fluctuations? This problem will be discussed in the last section.

6.3. Model

What type of a model is best suited for a cell to answer the question raised above? With all the current biochemical knowledge, we can say that one could write down several types of intended models. Due to the complexity of a cell, there is a tendency of building a complicated model in trying to capture the essence of a cell. However, doing so only makes one difficult to extract new concepts, although simulation of the model may produce similar phenomena as those in living cells. Therefore, to avoid such failures, it may be more appropriate to start with a simple model that encompasses only the essential factors of living cells. Simple models may not produce all the observed natural phenomena, but are comprehensive enough to bring us new thoughts on the course of events taken in nature.

In setting up a theoretical model here, we do not put many conditions to imitate the life process. Rather we impose the postulates as minimum as possible, and study universal properties in such system. For example, as a minimal condition for a cell, we consider a system consisting of chemicals

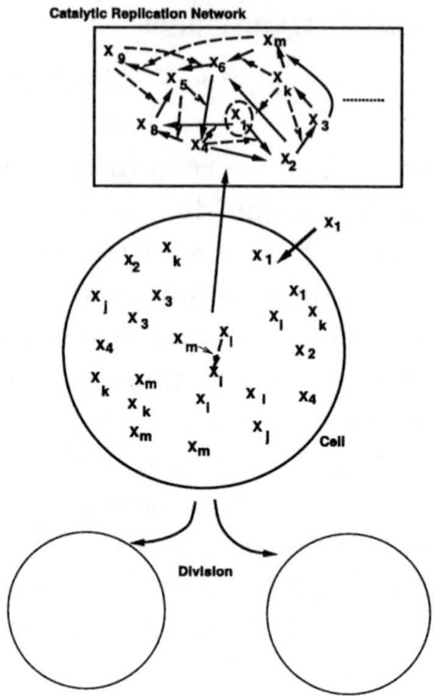

Fig. 6.1. Schematic representation of our modeling strategy of a cell.

separated by a membrane. The chemicals are synthesized through catalytic reactions, and accordingly the amount of chemicals increases, including the membrane component. As the volume of this system is larger, the surface tension for the membrane can no longer sustain the system, and it will divide. Under such minimum setup as will be discussed later, we study the condition for the recursive growth of a cell.

Let us start from simple argument for a biochemical process that a cell that grows must at least satisfy. A huge number of chemicals that catalyze each other is spatially arranged in a cell, and in some problems such spatial arrangement is very important, but as a minimal condition, let us discard the spatial configuration of molecules within a cell, since they are not rigidly fixed but can move around randomly to some degree. Hence, we consider just the composition of chemicals in a cell. These molecules change their number through reaction among these molecules. Since most reactions are catalyzed by some other molecules, the reaction dynamics consist of a catalytic reaction network.

Assuming that some reaction processes are fast, they can be adiabatically eliminated. Also, most of fast reversible reactions can be eliminated by assuming that they are already balanced. Then we need to discuss only the concentration (number) of molecules species, that change relatively slowly. For example by assuming that enzyme is synthesized and decomposed fast, the concentrations can be eliminated, to give catalytic reaction network dynamics consisting of the reactions with

$$X_i + X_j \rightarrow X_\ell + X_j \tag{3}$$

where X_j catalyzes the reaction[8,4]. If the catalysis progresses through several steps, this process is replace by

$$X_i + mX_j \rightarrow X_\ell + mX_j \tag{4}$$

leading to higher order catalysis[9].

Now, the internal state of the cell can be represented by a set of numbers $(n_1, n_2, \cdots, n_k)$, where n_i is the number of molecules of the chemical species X_i with i ranging from $i = 1$ to k. For the internal chemical reaction dynamics, we chose a catalytic network among these k chemical species, where each reaction from some chemical X_i to some other chemical X_j is assumed to be catalyzed by a third chemical X_ℓ, as depicted above. The rate of increase of n_j (and decrease of n_i) through this reaction is given by $\epsilon n_i n_\ell / N^2$, where ϵ is the coefficient for the chemical reaction and N is the total number of chemicals $N = \sum_i n_i$ in a cell, which is assumed to correspond to the cell volume. For simplicity all the reaction coefficients were chosen to be equal, and the connection paths of this catalytic network were chosen randomly such that the probability of any two chemicals X_i and X_j to be connected is given by the connection rate ρ.

For a cell to grow, some resource chemicals must be supplied through membrane, which are successively transformed to other chemicals through this catalytic reaction network. These resources (nutrients) are supplied from the environment by diffusion through the membrane (with a diffusion coefficient D). (Note that the nutrient chemicals have no catalytic activity, since they are not products by intra-cellular reactions. Indeed, there does not occur catalytic reactions in the environment.)Besides these nutrients, some of these chemicals may penetrate[b] the membrane and diffuse out while others will not. With the synthesis of the impenetrable chemicals that do not diffuse out, the total number of chemicals N in a cell, and accordingly the cell volume can increase. Then, when the cell volume is larger than some value, it is expected to divided. We study how this cell growth is sustained by dividing a cell into two when the volume is larger than a given threshold N_{max}. In the cell division process, the mother cell's molecules are evenly split among the two daughter cells, chosen randomly.

Summing up, the basic picture for a simple toy cell we take is given as in Fig. 6.1. In our numerical simulations, we randomly pick up a pair of molecules in a cell, and transform them according to the reaction network. In the same way, diffusion through the membrane is also computed by randomly choosing molecules inside the cell and nutrients in the environment. In the case with $N \gg k$ (i.e. continuous limit), the reaction dynamics is represented by the following rate equation:

$$
\begin{aligned}
dn_i/dt = &\sum_{j,\ell} Con(j,i,\ell)\,\epsilon\, n_j\, n_\ell/N^2 \\
&-\sum_{j',\ell'} Con(i,j',\ell')\,\epsilon\, n_i\, n_{\ell'}/N^2 + D\sigma_i(\overline{n_i}/V - n_i/N),
\end{aligned}
\tag{5}
$$

where $Con(i,j,\ell)$ is 1 if there is a reaction $i+\ell \to j+\ell$, and 0 otherwise, whereas σ_i takes 1 if the chemical i is penetrable, and 0 otherwise. The third term describes the transport of chemicals through the membrane, where $\overline{n_i}$ is a constant, representing the number of the i-th chemical species in the environment and V denotes the volume of the environment in units of the initial cell size. The number $\overline{n_i}$ is nonzero only for the nutrient chemicals.

Remark: Models for specific gene expression or signal transduction have been extensively studied these days, which are relevant to understand specific function of a cell. There, we can construct a more detailed model. In our study, we are interested in a recursive production of a 'whole cell', and such partial model is not adequate. Instead, we take a 'crude' catalytic reaction network model that at least include reproduction of the whole set of chemicals.

---

[b]Even if the reaction coefficient and diffusion coefficient of penetrating chemicals are not identical but distributed, the results reported here are obtained.

6.4. Zipf Law

If the total number of molecules N_{max} is larger than the number of chemical species k, the population ratios $\{n_i/N\}$ are generally fixed, since the daughter cells inherit the chemical compositions of their mother cells. For $k > N_{max}$, the population ratios do not settle down and can change from generation to generation. In both cases, depending on the membrane diffusion coefficient D, the characteristics of intra-cellular reaction dynamics change as will be discussed below. Equivalently, this change of intra-cellular dynamics also appear when changing the connection rate ρ.

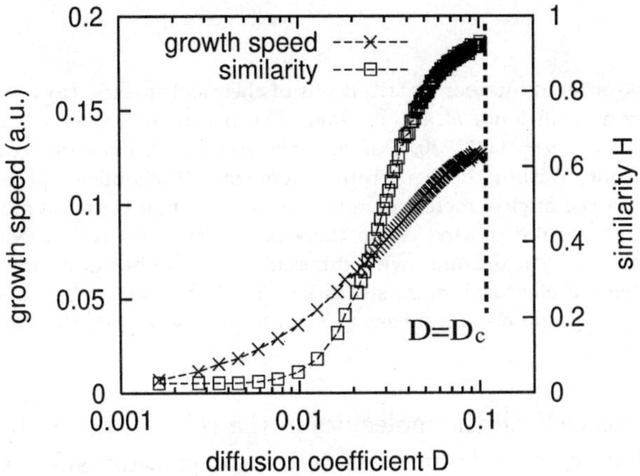

Fig. 6.2. The growth speed of a cell and the similarity between the chemical compositions of the mother and daughter cells, plotted as a function of the diffusion coefficient D. The growth speed is measured as the inverse of the time for a cell to divide. The degree of similarity between two different states m (mother) and d (daughter) is measured as the scalar product of k-dimensional vectors $H(\mathbf{n}_m, \mathbf{n}_d) = (\mathbf{n}_m/|\mathbf{n}_m|) \cdot (\mathbf{n}_d/|\mathbf{n}_d|)$, where $\mathbf{n} = (n_1, n_2, ..., n_k)$ represents the chemical composition of a cell and $|\mathbf{n}|$ is the norm of $\mathbf{n}$. Both the growth speed and the similarity are averaged over 500 cell divisions. Note that the case $H = 1$ indicates an identical chemical composition between the mother and daughter cells. Reproduced from [4].

As D is increased, the growth speed of a cell is increased as shown in Fig. 6.2, since the intake of nutrients is hastened. However, there is a critical value $D = D_c$ beyond which the cell cannot grow continuously. When $D > D_c$, the flow of nutrients from the environment is so fast that the internal reactions transforming them into chemicals sustaining 'metabolism' cannot

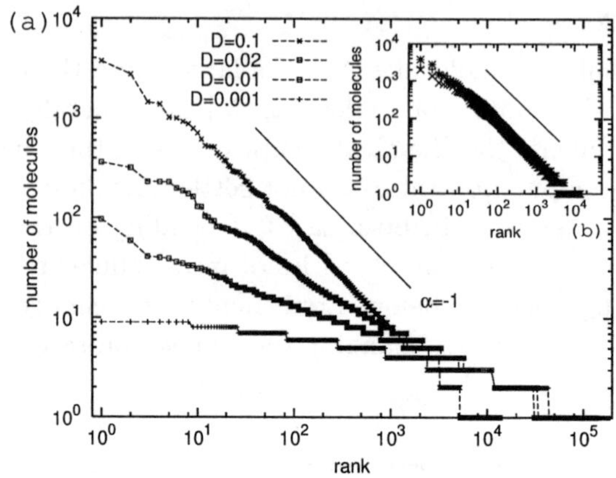

Fig. 6.3. Rank-ordered number distributions of chemical species. (a) Distributions with different diffusion coefficients D are overlaid. The parameters were set as $k = 5 \times 10^6$, $N_{max} = 5 \times 10^5$, and $\rho = 0.022$. 30 % of chemical species are penetrating the membrane, and others are not. Within the penetrable chemicals, 10 chemical species are continuously supplied to the environment, as nutrients. In this figure, the numbers of nutrient chemicals in a cell are not plotted. With these parameters, D_c is approximately 0.1. (b) Distributions at the critical points with different total number of chemicals k are overlaid. The numbers of chemicals were set as $k = 5 \times 10^4$, $k = 5 \times 10^5$, and $k = 5 \times 10^6$, respectively. Other parameters were set the same as those in (a). Reproduced from [4].

keep up. In this case all the molecules in the cell will finally be substituted by the nutrient chemicals and the cell stops growing since the nutrients alone cannot catalyze any reactions to generate impenetrable chemicals. Continuous cellular growth and successive divisions are possible only for $D \le D_c$. When the diffusion coefficient D is sufficiently small, the internal reactions progress faster than the flow of nutrients from the environment, and all the existing chemical species have small numbers of approximately the same level. As shown in Fig. 6.2, the growth speed of a cell is maximal at $D = D_c$. This suggests that a cell whose reaction dynamics are in the critical state should be selected by natural selection. Indeed, simulations with evolution of these cells support this[10].

Second, at the critical point, the similarity of chemical compositions between the mother and daughter cell is maximal as shown in Fig. 6.2. Indeed, for $k > N$, the chemical compositions differ significantly from generation to generation when $D \ll D_c$. When $D \approx D_c$, several semi-stable states with distinct chemical compositions appear. Daughter cells in the semi-stable states inherit chemical compositions that are nearly identical to their mother cells over many generations, until fluctuations in molecule numbers induce a transition to another semi-stable state. This means that the most faithful transfer of the information determining a cell's intracellular state is at the critical state. To characterize the similarity of cell compositions by divisions, we use the inner product of composition vectors of mother and daughter cells[4,11,7]. As plotted in Fig. 6.2, the similarity is maximal around $D \sim D_c$. In this state, chemical compositions of cells are well transferred to offspring through divisions, in the amidst of fluctuations. To sum up, the faithful reproduction is possible for a cell at $D \sim D_c$.

Now, we study the statistics of the abundances of chemicals focusing on the case with $D \sim D_c$. The rank-ordered number distributions of chemical species in our model are plotted in Fig. 6.3, where the ordinate indicates the number of molecules n_i and abscissa shows the rank determined by n_i. As shown in the figure, the slope in the rank-ordered number distribution increases with an increase of the diffusion coefficient D. We found that at the critical point $D = D_c$, the distribution converges to a power-law with an exponent -1. This type of power-law was first studied in linguistics as the frequency of appearance of words by Zipf[12], and is called Zipf law.

The power-law distribution at this critical point is maintained by a hierarchical organization of catalytic reactions, where the synthesis of higher ranking chemicals is catalyzed by lower ranking chemicals. For example, major chemical species are directly synthesized from nutrients and catalyzed by chemicals that are slightly less abundant. The latter chemicals are mostly synthesized from nutrients (or other major chemicals), and catalyzed by chemicals that are much less abundant. In turn these chemicals are catalyzed by chemicals that are even less abundant, and this hierarchy of catalytic reactions continues until it reaches the minor chemical species. In fact, in the case depicted in the inset of Fig. 6.3, a hierarchical organization of catalytic reactions with $5 \sim 6$ layers is observed at the critical point, as schematically shown in Fig. 6.4.

Based on this catalytic hierarchy, the observed exponent -1 can be explained using a mean field approximation. First, we replace the concentration n_i/N of each chemical i, except the nutrient chemicals, by a single average concentration (mean field) x, while the concentrations of nutrient chemicals S is given by the average concentration $S = 1 - k^*x$, where k^* is the number of non-nutrient chemical species. From this mean field equation, we obtain $S = \frac{DS_0}{D+\epsilon\rho}$ with $S_0 = \sum_j \overline{n_j}/V$. With linear stability analysis, the solution with $S \neq 1$ is stable if $D < \frac{\epsilon\rho}{S_0 - 1} \equiv D_c$. Indeed, this critical value does not differ much from numerical observation.

Next, we study how the concentrations of non-nutrient chemicals differentiate. Suppose that chemicals $\{i_0\}$ are synthesized directly from nutrients through catalyzation by chemicals j. As the next step of the mean-field approximation we assume the concentrations of the chemicals $\{i_0\}$ are larger than the others. Now we represent the dynamics by two mean-field concentrations; the concentration of $\{i_0\}$ chemicals, x_0, and the concentration of the others, x_1. The solution with $x_0 \neq x_1$ satisfies $x_0 \approx x_1/\rho$ at the critical point D_c. Since the fraction of the $\{i_0\}$ chemicals among the non-nutrient chemicals is ρ, the relative abundance of the chemicals $\{i_0\}$ is inversely proportional to this fraction. Similarly, one can compute the relative abundances of the chemicals of the next layer synthesized from i_0. At $D \approx D_c$, this hierarchy of the catalytic network is continued. In general a given layer of the hierarchy is defined by the chemicals whose synthesis from the nutrients is catalyzed by the layer one step down in the hierarchy. The abundance of chemical species in a given layer is $1/\rho$ times larger than chemicals in the layer one step down. Then, in the same way as this hierarchical organization of chemicals, the increase of chemical abundances and the decrease of number of chemical species are given by factors of $1/\rho$ and ρ, respectively. This is the reason for the emergence of power-law with an exponent -1 in the rank-ordered distribution. Within a given layer, a further hierarchy exists, which again leads to the Zipf rank distribution.

In general, as the flow of nutrients from the environment increases, the hierarchical catalyzation network pops up from random reaction networks. This hierarchy continues until it covers all chemicals, at $D \to D_c - 0$. Hence, the emergence of a power-law distribution of chemical abundances near the critical point is quite general, and does not rely on the details of our model, such as the network configuration or the kinetic rules of the reactions.

We have checked the universality of the Zipf law by adopting distributed network connectivity, and distributed parameters, and so forth. Our result is invariant against these changes of the model. This power law in abundances seems to be a quite universal law as long as a cell achieves a recursive growth (i.e., relatively faithful reproduction) and efficient growth. Since the current cell also satisfies relatively faithful reproduction (as well as effective growth for a suitable condition), the above behavior at $D \sim D_c$ may be expected to be true also in the present cell. Then, does this Zipf law hold in the present cell? In order to investigate possible universal properties of the reaction dynamics, examined are the distributions of the abundances of expressed genes (that are the abundances of the mRNA that produce corresponding proteins). As will be discussed later, the data from several tissues suggest the validity of the present law[4].

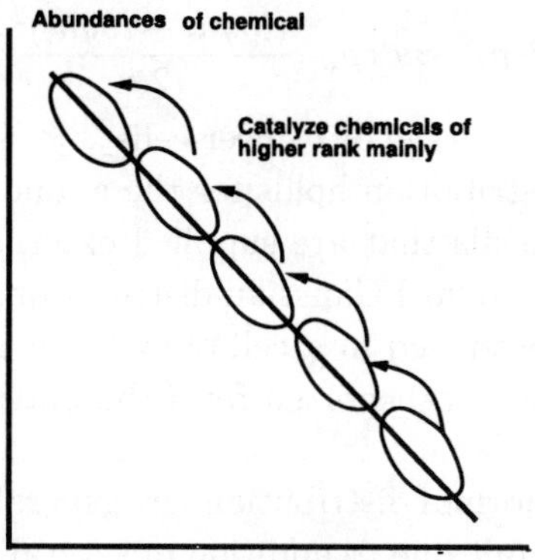

Fig. 6.4. Schematic representation of catalytic cascade. Higher rank chemicals are mainly catalyzed by a lower level. This is a schematic representation showing a rough structure and indeed there are several other cascade paths also.

Remark: Corresponding to the Zipf law in the rank-abundance relationship, the distribution $p(x)$ of the chemical species with abundance x is proportional to x^{-2}. This is easily obtained by noting that the rank distribution, i.e., the abundances x plotted by rank n can be transformed the density distribution function $p(x)$, which is the probability that the abundance is between x and $x + dx$. Recalling that $dx = dx/dn \times dn$, there are $|dx/dn|^{-1}$ chemical species between x and $x+dx$. Thus, if the abundance-rank relation is given by a power-law with exponent -1, $p(x) = |dx/dn|^{-1} \propto n^2 \propto x^{-2}$.

6.5. Log-Normal Distribution

In the lase subsection, we discussed the average abundances of each chemical, and studied the universal statistic over molecule species. Since the chemical reaction process is stochastic, each number of molecule differs by cells. For example, if x_3 in the last section is 3000, it can be 2631, 3203, 2903, 3611,... for each cell. Now we study the distribution of each molecule number, sampled over cells. In Fig. 6.5, we plotted the number distributions of several chemicals in the condition $D \sim D_c$. We measure the number of chemicals when a cell is divided into two. Since the total number of chemicals is constant when the cell is divided, the abundances of each chemical are just proportional to the concentration of each. As shown in Fig. 6.5, the distribution is fitted quite well by the log-normal distribution i.e.,

$$P(n_i) \approx exp(-\frac{(logn_i - log\overline{n_i})^2}{2\sigma}), \tag{6}$$

where $\overline{n_i}$ indicates the average of n_i over cells[6].

This log-normal distribution holds for the abundances of all chemicals, except for a few chemicals that are supplied externally to a cell as nutrients, which obey the standard Gaussian distribution. In other words, those molecules that are reproduced in a cell obey the log-normal distributions, while those that are just transported from the outside of a cell follow the normal distributions.

Why does the log-normal distribution law generally hold, in spite of the threat by the central limit due to addition of several fluctuation terms? We have discussed already the possible mechanism for the log-normal distribution in §2. However, there is slight difference here. In the discussion of §2, there is autocatalytic reaction process, as given by $dx_i/dt = cx_jx_i$, which leads to multiplicative stochastic process. In the present example, the reaction process is given by $dx_i/dt = cx_jx_\ell$. Thus, the discussion of §2 is not directly applied.

Still, there is a multiplicative reaction process here. Furthermore, as discussion in §3, there is a cascade reaction process, to support the recursive production around the critical state $D \sim D_c$. There, only a part of possible reaction pathways are used dominantly, to organize a cascade of catalytic reactions so that a chemical in the i-th group is catalyzed by the $(i+1)-th$, and that in the $(i+1)-th$ group is catalyzed by the $(i+2)-th$. A "modular structure" with groups of successive catalytic reactions is self-organized in the network. Then the fluctuations influence multiplicatively through the cascade. By taking logarithm of concentrations (i.e., $logx_m$), these successive multiplication are transformed to successive addition. Now, consider of the average of many random numbers. The distribution of the average approaches the Gaussian distribution, as given by the central limit theorem. The distribution of $logx_m$ is then expected to obey the Gaussian distribution. Hence, the log-normal distribution of x_m is derived. Note that, at the critical state we are concerned, this cascade of catalytic reaction continues for all chemical species that are reproduced, and the log-normal distribution holds clearly.

Next we discuss how the magnitude of fluctuations depend on each chemical species. As shown in Fig. 6.5, the width of the distribution does not change much independently of its average abundances, when plotted in the logarithmic scale. This suggests a relationship between the fluctuation and the average for all chemicals. We have thus plotted the standard deviation of each chemical $\sqrt{\overline{(n_i - \overline{n_i})^2}}$ as a function of the average. As shown in Fig. 6.2, the standard deviation (*not* the variance) increases linearly with the average.

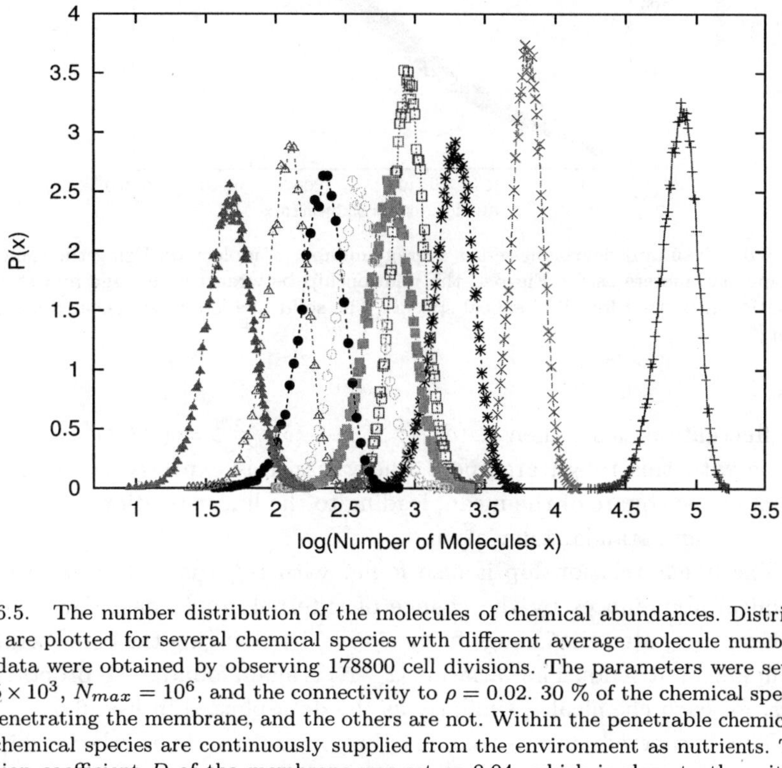

Fig. 6.5. The number distribution of the molecules of chemical abundances. Distributions are plotted for several chemical species with different average molecule numbers. The data were obtained by observing 178800 cell divisions. The parameters were set as $k = 5 \times 10^3$, $N_{max} = 10^6$, and the connectivity to $\rho = 0.02$. 30 % of the chemical species are penetrating the membrane, and the others are not. Within the penetrable chemicals, two chemical species are continuously supplied from the environment as nutrients. The diffusion coefficient D of the membrane was set as 0.04, which is close to the critical value D_c. Reproduced from [6].

To discuss the relationship between the mean and the standard-deviation, one should recall the steady growth and cascade structure in catalytic reactions. Consider two chemicals i and j, one of which (j) belongs to a group of catalyzing molecules for the other. Then the balance between the synthesis and conversion implies $x_i \times A - x_j \times B = 0$, where A and B are average concentrations of other chemicals involved in the catalytic reaction. Assuming the steady growth of a cell, the average concentration satisfies $\overline{x_i}/\overline{x_j} = A/B$. If we further assume the steady state condition for

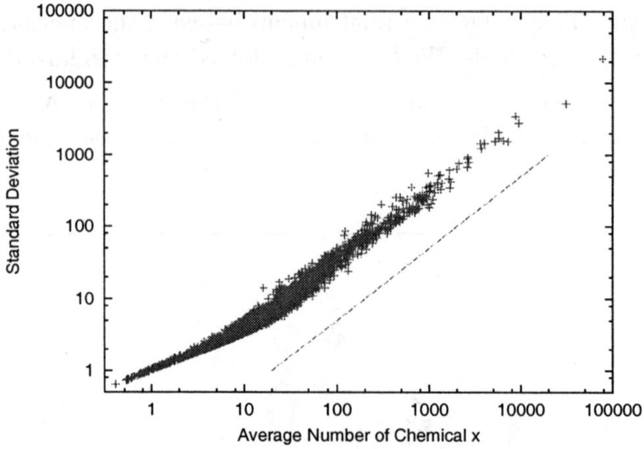

Fig. 6.6. Standard deviation versus average number of molecules. Using the same data set and parameters as for Fig 6.4, the relationship between the average and standard deviation is plotted for all chemical species. The solid line is for reference. Reproduced from [6].

the fluctuations also, then $< (\delta x_i)^2 > / < (\delta x_j)^2 >= (A/B)^2 = \overline{x_i}^2/\overline{x_j}^2$. Hence with this rough argument, the variance is expected to be proportional to the square of the mean, leading to the linear relationship between the mean and standard deviation.

The linear relationship is also found with regards to the variation of chemical abundances by the change of external conditions. For example, we have computed the change from $\overline{x_i}$ to $\overline{x'_i}$ by changing the concentrations of supplied nutrients. The variation $|\overline{x'_i}$ -$\overline{x_i}|$ is again found to be proportional to $\overline{x_i}$ for each chemical i, similarly as the data plotted in Fig. 6.6.

The discovered laws on the distribution and the liner relationship between the average and fluctuation are universally observed, near the critical point with the largest reproduction speed, hold generally and do not rely on the details of the model, such as the network configuration of the kinetic rules of the reactions, as has been confirmed from simulations of a variety of models.

Note, however, that the above arguments for the two laws are based on the steady growth of a cell with catalytic cascade process, realized at $D \sim D_c$. Indeed, as the parameter D is much smaller, all possible reaction pathways are used with a similar weight, where the cascades of catalytic reactions are replaced by random reaction network. In this case, the fluctuations of each molecule number are highly suppressed, and the distribution is close to normal Gaussian. The variance (not the standard deviation) increases linearly with the average concentrations. In other words, the behavior is "normal" as expected from the central limit theorem.

6.6. Experiment

6.6.1. *Confirmation of Zipf Law*

First we discuss the validity of Zipf law with regards to the abundances of many chemicals within the cell. In order to investigate possible universal properties of the reaction dynamics, we examined the distributions of the abundances of expressed genes, that are nothing but the abundances of the mRNA that produce corresponding proteins in a variety of organisms and a variety of tissues. In Ref.[4], they used the data publicly available from SAGE (Serial Analysis of Gene Expression) databases [14] over 6 organisms and more than 40 tissues. S AGE allows the number of copies of any given mRNA to be quantitatively evaluated by determining the abundances of the short sequence tags which uniquely identify it [15].

Following the numerical results of the model, we plotted the rank-ordered frequency distributions of the expressed genes, where the ordinate indicates the frequency of the observed sequence tags (i.e. the population ratio of the corresponding mRNA to the total mRNA), and the abscissa shows the rank determined from this frequency. As shown, the distributions follow a power-law with an exponent close to -1 (Zipf law). We observed this power-law distribution for all the available samples, including 18 human normal tissues, human cancer tissues, mouse embryonic stem cells, nematode (*C. elegans*), and yeast (*S.cerevisiae*) cells. All the data over 40 samples (except for 2 plant data) show the power-law distributions with the exponent in the range from $-1 \sim -0.9$. Even though there are some factors which may bias the results of the SAGE experiments, it seems rather unlikely that the distribution is an artifact of the experimental procedure. Indeed there are increasing supports for this Zipf law, by using a standard micro array.

6.6.1.1. *Confirmation of Laws on Fluctuations*

Now we discuss the confirmation of the laws of fluctuations. Recalling that the laws are expected to hold for the abundances of a protein synthesized within cells with recursive (steady) growth, we measured the distribution of the protein abundances in *Escherichia coli* that are in the log phase growth, i.e., in a stage of steady growth. To obtain the distribution of the protein abundances, we introduced the fluorescent proteins with appropriate promoters into the cells, and measured the fluorescence intensity by flow cytometry. To demonstrate the universality of the laws, we have carried out several sets of experiments by using a variety of promoters and also by changing places that the reporter genes are introduced (i.e., on the plasmid and on the genome).

Indeed, we have measured the distributions of the emitted fluorescence intensity from *Escherichia coli* cells with the reporter plasmids containing either (enhanced) green fluorescent protein. In general, the fluorescence intensity (the abundance of the protein) increases with the cell size. To avoid the effect of variation of cell size, which may also obey log-normal distribution, we normalized the fluorescence intensity by the volume of each cell, that is estimated by the forward-scatter (FS) signal from the flow cytometry. The distributions of this normalized fluorescence intensity are fitted well by log-normal, rather than Gaussian, distributions, even though each of the expressions is controlled by a different condition of the promoter. We have also measured the abundances of fluorescent protein expressed from the chromosome, which again is found to obey the log-normal distribution. It is furthermore interesting to note that the abundances of the fluorescent proteins, reported in the literature so far, are often plotted with a logarithmic scale[13].

It should be noted that the log-normal distribution of protein abundances is observed when the *Escherichia coli* are in the log phase of growth, i.e., when the bacteria are in steady growth. For other phases of growth, without steady growth, the distribution is found to be often deviated from the log-normal distribution. Note that the theory also supports the log-normal distribution for the steady growth case only, i.e., for a state with recursive production. If a cell is not in a stationary growth state but in a transient process switching from one steady state to another, the universal statistics can be violated.

6.7. Discussion

In the present chapter we have observed ubiquity of log-normal distribution, in several models. The fluctuations in such distribution are generally very large. This is in contrast to our naive impression that a process in a cell system must be well controlled.

Then, is there some relevance of such large fluctuations to biology? Quite recently, we have extended the idea of fluctuation-dissipation theorem in statistical physics to evolution, and proposed a linear relationship (or high correlation) between (genetic) evolution speed and (phenotypic) fluctuations. This proposition turns out to be supported by experimental data on the evolution of E Coli to enhance the fluorescence in its proteins[16]. Furthermore this phenotypic fluctuation is shown to be tightly related with the genetic variance, measured through phenotype[18]. Hence the fluctuations are important biologically.

The log-normal distribution is also rather universal in the present cell, as demonstrated in the distribution of some proteins, measured by the degree of fluorescence. We have to be cautious here, since too much strongly universal laws may not be so relevant to biological function. In fact, chemicals that obey the log-normal distribution may have too large fluctuations to control some function. For example, the abundances of DNA should be deviated from the log-normal distribution. Some other mechanism to suppress the fluctuation may work in a cell[c].

Note that in a system consisting of molecules synthesized with mutual catalysis, there can appear some key-stone chemicals. These key-stone molecules work as a controlling part. As discussed in Ref.[7], this key-stone molecule has a higher connection in reaction network. Then the increase of number fluctuations of such molecule result in a drastic change in the number of other molecules. Accordingly, to have stable recursive production of a cell, such key-stone molecules must have relatively smaller fluctuations, in contrast to the large fluctuations of other molecules arisen from the log-normal distribution. With a combination of such key-stone chemicals, information on recursive growth is generated, that works as a controller for synchronized growth.

Then, what mechanism is a candidate to decrease the fluctuation leading to deviation from log-normal distribution? In relationship with the above key-stone molecules, we previously proposed a minority control mechanism to suppress the fluctuation[19]. In a reproducing cell consisting of mutually catalytic molecules, those in minority have tendency to control the behavior of the cell and are preserved relatively well. These minority molecules are expected to play the role of key-stone molecule species.

---

[c]It is interesting to note that the weight distribution of adult human obeys the log-normal distribution, while the height distribution obeys the Gaussian distribution.

The next standard mechanism is negative feedback process. In general, the negative feedback can suppress the response as well as the fluctuation. Still, it is not a trivial question how chemical reaction can give rise to suppression of fluctuation, since to realize the negative feedback in chemical reaction, production of some molecules is necessary, which may further add fluctuations.

The other possible mechanism is the use of multiple parallel reaction paths. If several processes work sequentially, the fluctuations would generally be increased. When reaction processes work in parallel for some species, the population change of such molecule is influenced by several fluctuation terms added in parallel. If a synthesis (or decomposition) of some chemical species is a result of the average of these processes working in parallel, the fluctuation around this average can be decreased by the law of large numbers. Suppression of fluctuation by multiple parallel paths may be a strategy adopted in a cell.

In fact, the minority molecule species in a core catalytic network discussed in Ref.[7] has higher reaction paths and has relatively lower fluctuations. This is also consistent with the scenario that more and more molecules are related with the minority species through evolution, as discussed in [19]. With the increase of the paths connected with the minority molecules, the fluctuation of minority molecules is reduced, which further reinforces the minority control mechanism. Hence the increase of the reaction paths connected with the minority molecule species through evolution, decrease of the fluctuation in the population of minority molecules, and enhancement of minority control reinforce each other. With this regards, search for molecules that deviate from log-normal distribution should be important, in future. Here it is important to measure the distribution of chemicals in relationship with its characteristics (such as the connectivity) in the reaction network.

In physics, we are often interested in some quantities that deviate from Gaussian (normal) distribution, since the deviation is exceptional. Indeed, in physics, search for power-law distribution or log-normal distributions has been popular over a few decades, because they are exceptional. On the other hand, a biological unit can grow and reproduce, to increase the number. For such system, the components within have to be synthesized, so that amplification process is common. Then, the fluctuation is also amplified. In such system, the power-law or log-normal distributions are quite common, as already discussed here, and as is also shown in several models and experiments [4,6]. In this case, the Gaussian (normal) distribution is not so common. Then exceptional molecules that obey the normal distribution with regards to their concentration may be more important.

Note that our findings are not restricted to a cell. As long as a system grows, through production by mutual catalytic processes, universal properties we discussed here will generally appear. Society of human beings with economic production process is such an example, and indeed Zipf law was first studied in human activities[12]. For example, such power law is also observed in the wealth distribution. Universality in a reproduction system revealed in cell biology will be relevant to social dynamics. Indeed, formation of hierarchy, minority control, as well as differentiation that was discussed in reproducing cell models[4,8,9,19] provide conceptual tools to understand history, the dynamics of human society[21]. Furthermore, industry production process, a key issue in the present proceedings, is also such an example of growth system with catalytic process[22]. As discussed in the present volume, the industrial production system is also in the amidst of large fluctuations, and suppression or control of such fluctuations should be important. Ubiquity of log-normal distributions, as well as control mechanisms discussed here may be relevant to such problems.

References

1. M. Eigen and P. Schuster, *The Hypercycle* (Springer, 1979).
2. F. Dyson, *Origins of Life*, Cambridge Univ. Press., 1985
3. S.A. Kauffman, *The Origin of Order*, Oxford Univ. Press. 1993
4. C. Furusawa and K. Kaneko, Phys. Rev. Lett. 90 (2003) 088102
5. Elowitz, M. B., Levine, A. J., Siggia, E. D. & Swain, P. S. (2002) Science 297, 1183
6. C. Furusawa, T. Suzuki, A. Kashiwagi, T. Yomo and K. Kaneko ; Ubiquity of Log-normal Distribution in gene expression, BIOPHYSICS 1 (2005) 25
7. K. Kaneko, *Phys. Rev. E* **68** 031909 (2003)
8. K. Kaneko and T. Yomo, B. Math.Biol. 59 (1997) 139 J. Theor. Biol., 199 243-256 (1999)
9. Furusawa C. & Kaneko K., Bull.Math.Biol. 60; 659-687 (1998); Phys Rev Lett. 84:6130-6133 J. Theor. Biol. 209 (2001) 395-416; Anatomical Record, 268 (2002) 327-342
10. C. Furusawa and K. Kaneko, "Evolutionary origin of power-laws in Biochemical Reaction Network; embedding abundance distribution into topology", Phys. Rev. E (q-bio.PE/0504032)
11. D. Segré, D. Ben-Eli, D. Lancet, Proc. Natl. Acad. Sci. USA 97 (2000)4112; D. Segré et al., J. theor. Biol. **213** (2001) 481
12. G. K. Zipf, *Human Behavior and the Principle of Least Effort* (Addison-Wesley, Cambridge, 1949).
13. W. J. Blake, et al. *Nature* **422**, 633 (2003)

14. A.E. Lash et al., *Genome Research* **10**(7), 1051 (2000). V.E. Velculescu et al., *Cell* **88**, 243 (1997): S. J. Jones et al., *Genome Res.* **11**(8), 1346 (2001):
15. V.E. Velculescu et al. *Science* **270**, 484 (1995).
16. K. Sato, Y. Ito, T. Yomo, and K. Kaneko; Proc. Nat. Acad. Sci. USA 100 (2003) 14086-14090
17. Ito, Y., Kawama, T., Urabe, I. & Yomo, T., J. Mol. Evol. 58 (2004) 196
18. K. Kaneko and C. Furusawa, "An Evolutionary Relationship between Genetic Variation and Phenotypic Fluctuation", submitted to Proc. Roy. Soc. London .
19. Kaneko K, Yomo T. J. Theor. Biol. 312 (2002) 563-576
20. Matsuura T., Yomo T., Yamaguchi M, Shibuya N., Ko-Mitamura E.P., Shima Y., and Urabe I. Proc. Nat. Acad. Sci. USA 99 (2002) 7514-7517
21. K. Kaneko and A. Yasutomi, "Historical Science in view of Complex Systems Studies", in *Current Status of Economics*, ed. M. Yoashida, in Japanese, 2005
22. A. Ponzi, A. Yasutomi and K. Kaneko, "Economic Cycles in an Evolving Economic Production Network", JEBO, in press

CHAPTER 7

INTRACELLULAR NETWORKS OF INTERACTING MOLECULAR MACHINES

Alexander S. Mikhailov

Abteilung Physikalische Chemie, Fritz-Haber-Institut der Max-Planck-Gesellschaft

Faradayweg 4-6, D-14195 Berlin, Germany

E-mail: mikhailov@fhi-berlin.mpg.de

Biological cells can be viewed as networks formed by molecular machines. The coordination in operations of individual machines in such networks is not externally imposed, but develops internally as a result of interactions between the machines, through a self-organization process. As an example, we consider a population of allosteric enzymes with product activation and show that this network can undergo spontaneous transitions to different synchronous regimes corresponding to just-in-time production. Analyzing a toy model of cross-coupled neural networks, we find furthermore that similar synchronization transitions are possible when the dynamics of individual elements is highly complex.

7.1. Introdution

A living cell is a tiny chemical reactor where tens of thousands of chemical reactions can simultaneously go on. The very fact that these reactions proceed in a regular and predictable manner, despite thermal fluctuations and variations in environmental conditions, already indicates a high degree of organization in this system. The biochemical activity of a cell can be compared with operation of a large industrial factory where certain parts are produced by a system of machines[1]. Products of one machine are then used by other machines for manufacturing of their products or for regulation of their functions.

Two possible modes of operation of such a factory can be imagined. In the asynchronous mode, the parts produced by all machines are first deposited and accumulated in a common store. They are taken back from this store by other machines, when the parts are needed for further production. This kind of organization is not however optimal, since it requires large storage facilities and many transactions. It becomes deficient when intermediate products are potentially unstable and can easily be lost or damaged during the storage process.

When the synchronous operation mode is employed, the intermediate products, required for a certain operation step in a given machine, are released by other machines and become available exactly at the moment when they are needed. Hence, large storage facilities are eliminated and the entire process may run much faster. Such synchronous operation mode is also known as "just-in-time production".

For a living cell, the role of the 'machines' is played by individual enzyme molecules. The 'parts' are intermediate product molecules, which may also serve to allosterically regulate the catalytic activity of other enzymes. Under certain conditions, the biochemical subsystems of a cell may operate in the synchronous mode. When this occurs, the entire population of reacting and interacting molecules can be viewed as a highly connected dynamic molecular network.

The synchronous manufacturing process implies much more complicated management than the asynchronous operation mode. In a real factory, this is achieved by careful planning and detailed external control of the production. Such a rigid external control at a molecular level inside a cell is however impossible. The coherence, underlying the synchronous production, emerges in this case as a natural consequence of interactions between individual elements, i.e. it represents a special kind of self-organization phenomena. Remarkably, the predictable coherent operation of the complex molecular machinery must be maintained under very stringent conditions, when strong thermal fluctuations are present and the parts needed in the assembly lines of the cell are often transported just by random Brownian motion of molecules.

If the asynchronous mode is employed, functioning of an industrial factory is well described by a scheme indicating involved machines, intermediate products, and performed operations. If, furthermore, the operation rates of the machines and the rate of supply of the initial raw materials are known, the efficiency of the factory, i.e. its overall production rate, can easily be estimated.

In contrast to this, fine dynamical coordination of different operations in the synchronous mode with just-in-time production requires much more detailed information about properties of individual machines and their operation cycles. Such system parameters as the time needed for transportation and delivery of intermediate products and the duration of a single machine cycle become significant. Moreover, the mechanistic details of individual machine cycles, such as, for example, the moment inside a cycle when the product is released and the duration of the subsequent recovery phase of the machine leading to full restoration of its operation capacity, are already important.

For a biochemical system, the asynchronous operation mode corresponds to the classical kinetic regime. This regime can indeed be completely characterized by its reaction scheme and by the information about rate constants of elementary reaction steps. The above arguments show, however, that the knowledge of the reaction scheme and of the rate constants would not be sufficient to describe non-classical synchronous kinetics. When this different regime is realized, properties of molecular turnover cycles in individual enzyme molecules become equally essential in determining the system performance.

The self-organization processes, leading to emergence of coherent molecular networks, are different from the self-organization phenomena in classical reaction-diffusion systems. The latter typically lead to the appearance of complex spatiotemporal patterns or to the development of slow kinetic oscillations[2,3]. They are not, however, characterized by the presence of rigid microscopic correlations between individual molecular reaction events which represent a distinctive feature of a molecular network.

Indeed, slow oscillations may also develop in the considered above factory operating in the asynchronous regime. They would then mean that, as a result of a dynamic instability, the amounts of various intermediate parts in the store are slowly changing with time, on the characteristic temporal scales which are much longer than the duration of operation cycles of individual involved machines. The appearance of such oscillations does not, of course, influence the principal functional organization of the production process and does not lead to rigid correlations between individual machine cycles.

Emergence and functioning of molecular networks refers, therefore, to a different form of self-organization which is expressed at a level of correlations and synchrony in terms of individual molecular dynamics of a reacting population of molecules.

7.2. Networks of Protein Machines

The conditions for microscopic self-organization of chemical reactions in small spatial volumes are determined by relations between principal time scales of the involved physical and chemical processes. Estimates of these time scales can be obtained by considering a biochemical system formed by enzyme molecules whose activity is allosterically regulated by intermediate product molecules of a smaller molecular weight.To obtain such estimates [4,5], we assume that the reaction proceeds in a small compartment representing a single globular volume filled with liquid. The molecules inside the compartment perform random diffusive motion described by the Fick law.

The mixing time, t_{mix}, is defined as the time after which a regulatory molecule, released at some point in the volume, can be found with equal probability anywhere inside it. If the volume has the linear size L and the diffusion constant of the regulatory molecules is D, the mixing time can be estimated, in the order of magnitude, as

$$t_{mix} = \frac{L^2}{D}. \tag{1}$$

Another important characteristic time of the process is the traffic time, $t_{traffic}$, which is defined as follows: suppose that we have released a regulatory particle somewhere inside the volume that contains only one target enzyme molecule. Then ttraffic represents the characteristic time after which the regulatory particle will find this enzyme molecule and bind itself to the atomic target group on its surface.

It should be noted that the last stage of this process could actually be quite complicated and involve docking by electrostatic interactions and two-dimensional diffusion of the regulatory molecule over the surface of the enzyme until the target site is reached[6]. In our simple treatment we neglect all these possible complications and assume that the regulatory molecule performs free Brownian motion in the volume until it touches a small target of radius R attached to the surface of the enzyme molecule. Once touching has occurred, the regulatory molecule becomes bound to the target site.

Because the molecular weight of enzymes is larger than that of the regulatory molecules, the targets can be viewed as immobile. Under these conditions, the characteristic traffic time can be estimated using the concepts of the theory of diffusion-controlled reactions[7],[8]. In the order of magnitude, it is given by[9,5]

$$t_{traffic} = \frac{L^3}{DR}, \tag{2}$$

where R is the radius of a target (if the size of the atomic target group in the enzyme is comparable to the size of a regulatory molecule, it should be replaced by the sum of the two respective radii). The traffic time is related to the mixing time as $t_{traffic}\tilde{}(L/R)t_{mix}$.

To obtain numerical estimates of mixing and traffic times, we take $R = 1 - 10$ nm as the characteristic size of the atomic target group and $D = 10^{-5} - 10^{-6}$ cm^2/s as the diffusion constant of the regulatory molecules. For a compartment of size $L = 1$ mm this yields the mixing time $t_{mix} = 1 - 10$ ms and the traffic time $t_{traffic} = 0.1 - 10$ s. For smaller compartments of size $L = 0.1$ mm the estimates are $t_{mix} = 0.01 - 0.1$ ms and $t_{traffic} = 0.1 - 10$ ms. We see that the traffic time depends strongly on the compartment size (as the cube of L). This high sensitivity on the volume size already indicates that one should expect very special kinetic regimes inside small cells and cellular compartments.

The traffic time shows the time needed for a regulatory molecule to find a given molecular target. If the volume contains N such targets and they are independently and randomly distributed inside it, the transit time needed by a regulatory molecule to find one of the targets can be estimated as $t_{transit} = (1/N)t_{traffic}$ or explicitly as:

$$t_{transit} = \frac{L3}{NDR}. \tag{3}$$

Note that $L_{corr} = (Dt_{transit})^{1/2}$ yields the mean distance passed by the regulatory molecule before it finds a target and this characteristic length can be considered as the correlation radius of the reaction. This distance can be compared with the linear size L of the reaction volume.

When $L \gg L_{corr}$, the target will be found in a close proximity of the point where the regulatory product molecule has been released. Since the regulatory molecules, conveying information about the reaction events, are trapped in this case not far from the points where they have been produced, they cannot maintain long-range correlations in the reacting system.

A completely different situation is found if the condition $L < L_{corr}$ is satisfied. Now, the first target is found traveled extensively through the reaction volume and has crossed it many times. It means that the regulatory molecule would be able, with equal probability, to find any of N enzymes in the population, no matter where they are located inside the volume. The latter condition can also be expressed in terms of the characteristic times of the diffusion problem, i.e. can be written as:

$$t_{mix} \ll t_{transit}. \tag{4}$$

When the transit time is much larger than the mixing time, the first target is found with equal probability anywhere in the volume.

According to Eq. (3), the transit time is inversely proportional to the total number N of enzyme molecules. Therefore, the condition (4) can be satisfied only for sufficiently small numbers of the enzymes. By putting $t_{mix} = t_{transit}$, we derive the estimate for the critical number of the enzyme molecules:

$$N_{cr} = \frac{L}{R}. \tag{5}$$

Thus, it is controlled in the order of magnitude only by two parameters: the linear size L of the reaction volume and the radius R of the atomic target group on the surface of the enzyme molecule. When the number N of enzyme molecules is less than N_{cr}, the product-mediated allosteric interactions between enzymes extend over the entire reaction volume. Note that Eq. (5) yields the critical enzyme concentration

$$c_{cr} = \frac{1}{L^2 R}, \tag{6}$$

which decreases with the linear size L of the volume.

Taking $R = 1 - 10$ nm as the characteristic size of an enzyme target group, we see that the critical number of enzyme molecules in the volume of size $L = 1$ mm is $N_{cr} = 100 - 1000$, which corresponds to the enzymic concentrations of $c_{cr} = 10^{-7} - 10^{-6}$ M. For a smaller compartment of size $L = 0.1$ mm we obtain $N_{cr} = 10 - 100$ and $c_{cr} = 10^{-5} - 10^{-4}$ M.

The above characteristic diffusion times can be compared with the duration τ_0 of a single catalytic molecular cycle. This time has not been directly measured but some estimates of its magnitude could be obtained by measuring the maximal turnover rate of an enzymic reaction. This rate, reached under the condition of the substrate saturation, is controlled only by the molecular turnover. Hence, τ_0 can be roughly estimated as the inverse of the maximal turnover rate. Thus derived, the estimates for τ_0 may range, depending on a particular enzyme, from as small as fractions of a microsecond to several seconds[10]. If we consider relatively slow enzymes, the typical values would lie in the interval $\tau_0 = 10 - 100$ ms.

These characteristic microscopic molecular times can be compared with the characteristic diffusion times of the reaction. For the reactions in micrometer and submicrometer volumes involving hundreds or thousands of enzyme molecules this time may be much larger than both the mixing and the transit times in the reaction volume,

$$\tau_0 \ll \tau_{transit} \ll \tau_{mix}. \tag{7}$$

This is a remarkable result[5,12]: In a macroscopic system, all characteristic kinetic times of a chemical reaction are usually much longer than the duration of a single molecular reaction event and therefore these events can be treated as instantaneous. This forms the basis of the traditional kinetic theory formulated in terms of the Markov random processes. We see, however, that in very small spatial volumes for enzymic chemical reactions the opposite limit may be realized, so that the characteristic times of internal molecular dynamics of the enzymes become larger than the kinetic times of the reaction. When this occurs, the theoretical description of a chemical reaction should be essentially modified.

The considered system can be viewed as a population of active macromolecules that operate like molecular machines. Each internal molecular cycle of an enzyme results in the conversion of a substrate molecule into a molecule of the product. The cycle represents a sequence of conformation changes that can be understood as continuous motion along an internal reaction coordinate [11]. For allosteric enzymes, the molecular cycles can be externally controlled by binding of regulatory molecules. Generally, arrival of a regulatory molecule can lead either to acceleration or slowing down at some stages of a molecular cycle.

When regulatory molecules are produced by enzymes in the same population, this leads to interactions (or communication) between the cycles of different enzyme molecules. If Eq. (7) holds, the characteristic transport times of regulatory molecules between the enzymes are short as compared with the molecular cycle duration. Furthermore, if the transit time is larger than the mixing time in the volume, a regulatory molecule released by a given enzyme can influence, with equal probability, the catalytic cycle of any other enzyme in the population.

In the presence of strong allosteric interactions, the internal dynamics of individual enzyme molecules in such a network can be so significantly influenced by the dynamics of other enzymes that the whole system would perform coherent collective evolution in the multidimensional phase space of internal reaction coordinates, i.e. behave as a single dynamical object. This coherence would be manifested in rigid correlations between the individual catalytic cycles and the moments when the product molecules are released by different enzymes.

7.3. Coherent Molecular Dynamics

To show the possibility of a transition to coherent collective behavior of an enzymic population[14,15] we consider the reaction

$$S + E \to ES \to E + P, P \to 0 \tag{8}$$

where binding of substrate S by enzyme E is allosterically activated by product P. If the substrate concentration is kept constant, this reaction does not exhibit macroscopic rate oscillations. However, in small spatial volumes for sufficiently high intensities of the allosteric regulation periodic spiking of the catalytic activity of the whole enzymic population can develop. This spiking is a result of the synchronization of individual molecular cycles and thus represents an effect of coherent collective molecular dynamics.

The molecular mechanism of the reaction (8) includes several stages. An enzyme molecule E binds a molecule S of the substrate and the substrateenzyme complex ES begins a sequence of conformational changes that ends, after time τ_1, in the release of the product molecule P . When the product molecule has left, the conformation of the enzyme molecule returns to the initial state. This recovery stage takes time τ_2 during which binding of a substrate molecule cannot occur. Hence, only after the time $\tau_0 = \tau_1 + \tau_2$, which represents the duration of a molecular catalytic cycle, the enzyme molecule can begin a new cycle. Note that thus the molecular cycle has much in common with the sequence of transitions in an excitable neuron.

Figure 7.1 schematically shows an individual enzymatic molecular cycle. The small black circle represents the enzyme in the initial ground state, the light gray circle indicates the enzyme that has bound a substrate molecule and goes through a sequence of changes that would lead to the release of the product molecule. The enzyme releasing the product molecule is represented by a white circle. The enzyme going through the subsequent recovery stage is shown as a dark gray circle.

The initial binding of a substrate molecule is allosterically regulated. We assume that the probability of binding of a substrate by an enzyme

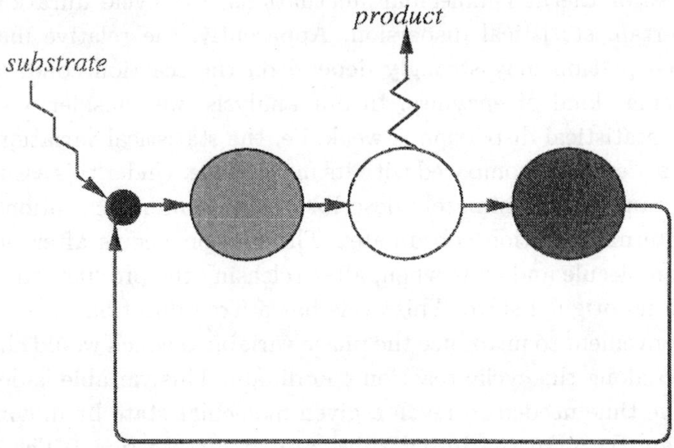

Fig. 7.1. Schematic view of a single molecular turnover cycle. The black dot indicates the enzyme molecule in the initial free state. The light gray circle depicts a sequence of states corresponding to the substrate-enzyme complex (this sequence has the duration τ_1); the product is released from the state shown as the white circle; the dark gray circle represents a sequence of recovery states (of duration τ_2) leading to the return of the enzyme to its initial state.

in its resting conformation is small. It becomes however greatly increased if an additional regulatory molecule is bound to a different site on the surface of the enzyme and this, by means of cooperative interactions within the enzyme macromolecule, facilitates the binding of the substrate. It is assumed that the enzyme has only one binding site for a regulatory molecule and its binding can only take place when an enzyme is in its resting state, i.e. when the previous catalytic cycle is finished. The regulatory molecule departs from the enzyme when the catalytic cycle begins.

In the considered hypothetical reaction (8), the regulatory role is played by the product molecules and thus the reaction is effectively autocatalytic. The product molecules decay and therefore their lifetimes are finite. To exclude immediate triggering of a new molecular catalytic cycle by the product molecule that has just been released by the enzyme, we assume that the lifetime of a product molecule is shorter than the molecular recovery time τ_2. Under this condition, the product molecules can usually execute their regulatory roles only by traveling through the volume and initiating the catalytic cycles of other enzymes in the population. If γ is the decay rate of product molecules, this condition can be expressed as $\gamma\tau_2 \gg 1$.

Because of thermal molecular fluctuations, the cycle duration would have a certain statistical dispersion. Apparently, the relative magnitude of this dis- persion may strongly depend on the reaction conditions and the particular kind of enzymes. In our analysis, we consider a situation when this statistical dispersion is weak, i.e. the statistical variations of the period τ_0 are small as compared with its mean value. Under this assumption, the cycle can be approximately described as deterministic motion along a certain internal 'reaction coordinate'. The motion begins after binding a substrate molecule and ends when, after releasing the product, the enzyme returns to its original state. This cycle has a fixed duration τ_0.

It is convenient to introduce the phase variable ϕ which would characterize motion along the cyclic reaction coordinate. This variable is defined as the relative time needed to reach a given molecular state by moving along the reaction path. In the initial substrate- free state $\phi = 0$.The product molecule is released at the state with the phase $\phi = \tau_1/\tau_0$. The cycle ends when the phase reaches the value $\phi = (\tau_1 + \tau_2)/\tau_0 = 1$.

Using this concept, the dynamics of a single enzyme molecule can be described as operation of an automaton, similar to that used by Wiener and Rosenblueth in their modeling of individual neurons. The model [14,15] includes the parameters ν_0 and ν_1 that characterize binding rates of a substrate molecule by a single molecule of the enzyme. The parameter ν_0 determines the probability per unit time of spontaneous non-activated binding of a substrate molecule. The parameter ν_1 represents the probability that, if the compartment contains a single regulatory product molecule, this molecule would lead per unit time to an activated binding of the substrate molecule by a given enzyme molecule.

For convenience, the time is divided in this model into small discrete steps Δt. The probabilities per a single time step of non-activated and activated substrate binding events are therefore been given by $w_0 = \nu_0 \Delta t$ and $w_1 = \nu_1 \Delta t$. The decay probability of a product molecule per a single time step is $g = \gamma \Delta t$.

In the discrete approximation, an individual cycle consists of $K = \tau_0/\Delta t$ steps and the product molecules are released after $K_1 = \tau_1/\Delta t$ time steps from the cycle initiation. Each automaton corresponds to a single enzyme and has K internal states described by the integer phase variable Φ. The state of the automaton i at the next discrete time moment T is determined by the algorithm:

$$\Phi_i(T+1) = \begin{cases} \Phi_i(T), \text{ if } 0 < \Phi_i(T) < K \\ 0, \text{ if } \Phi_i(T) = K \\ 1, \text{ with probability } w(m), \text{ if } \Phi_i(T) = 0 \\ 0, \text{ with probability } 1 - w(m), \text{ if } \Phi_i(T) = 0 \end{cases} \tag{9}$$

The probability w of cycle initiation depends on the number m of product molecules in the reaction volume:

$$w(m) = 1 - (1 - w_0)(1 - w_1)^m. \tag{10}$$

The products decay with the probability g per time step. The probability $\pi(m, m')$ that m' out of m product molecules decay within a single time step is therefore given by the binomial distribution:

$$\pi(m, m') = \frac{m!}{m'!(m - m')!} g^{m'} (1 - g)^{m - m'}. \tag{11}$$

Taking into account both production and decay processes, the number of product molecules in the reaction volume at the next time step is

$$m(T+1) = m(T) + \sum_{i=1}^{N} \delta(\Phi_i(T) - K_1) - m' \tag{12}$$

where N is the total number of enzymes. The number m' of disappearing product molecules is random; it obeys the probability distribution (11) where $m = m(T)$.

Below we present typical results[15] obtained using this model under different assumptions about the intensity of allosteric regulation which is specified by the parameter ν_1. As the initial condition in our simulations, a state with random distributions over the cycle phases and a small number of the product molecules is always chosen. The population consists of $N = 400$ enzyme molecules.

Figure 7.2 illustrates the kinetic regime of the reaction (8) when the intensity of allosteric regulation is low and therefore the individual molecular cycles are not correlated. The upper part of Fig. 7.2a displays the individual activity of 14 enzymes which have been at random selected among the entire population. The notations correspond to those used in Fig. 7.1. The segments of black lines indicate the intervals of time during which a given enzyme has been found in its initial state, waiting for the substrate binding event. The elongated gray-white-black graphic elements show the molecular catalytic cycles: the light gray color indicates the molecular states of

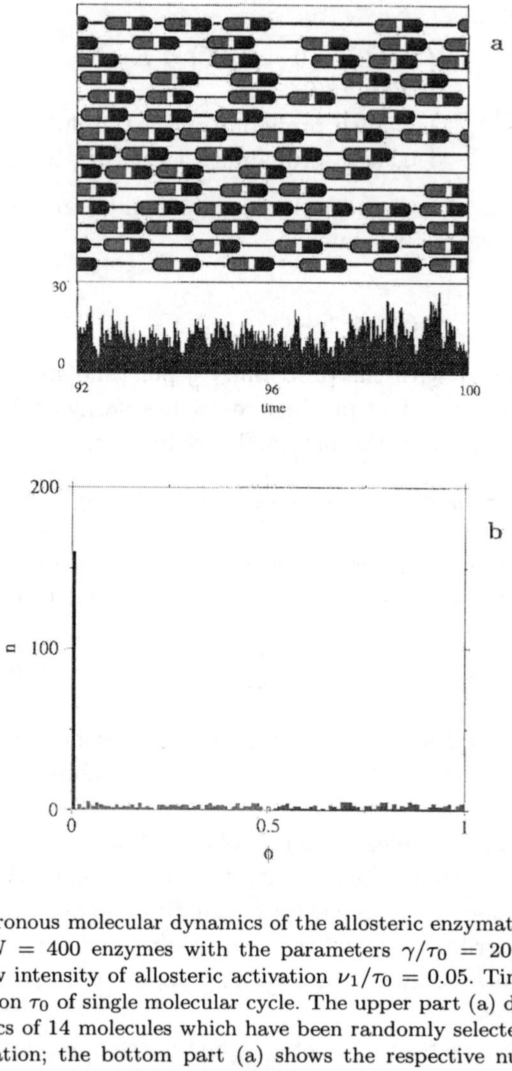

Fig. 7.2. Asynchronous molecular dynamics of the allosteric enzymatic reaction (8) for a population of $N = 400$ enzymes with the parameters $\gamma/\tau_0 = 20$, $\nu_0/\tau_0 = 1$, and $\tau_1/\tau_0 = 0.5$ at low intensity of allosteric activation $\nu_1/\tau_0 = 0.05$. Time is measured in units of the duration τ_0 of single molecular cycle. The upper part (a) displays individual molecular dynamics of 14 molecules which have been randomly selected from the whole interacting population; the bottom part (a) shows the respective number of product molecules in the compartment as function of time for this reaction (time is measured in units of the duration τ_0 of single molecular cycle). The histogram (b) displays the typical distribution of enzyme molecules over different cycle phases at a fixed time; the black peak at $\phi = 0$ shows the number of enzyme molecules in the free initial state, waiting to bind a substrate.

the substrate-enzyme complex, the white color shows the state when the product molecule is being released, and the dark gray color shows motion through the recovery states, ending with the return of the molecule to its initial state. The bottom part of Fig. 7.2a shows the total number of the product molecules in the compartment as function of time, which corresponds to the above-presented molecular enzymic dynamics. The time is measured in units of the duration of a single molecular cycle τ_0.

Figure 7.2b displays the histogram of the respective distribution of the entire enzymic population over different cycle phases, i.e. the number of enzyme molecules which are found at a fixed time moment in the molecular states corresponding to various values of the phase variable ϕ. The same color scheme as in Fig. 7.2a is used here to characterize different enzymic cycle stages.

We see that, as can be expected in this case, the cycles of individual enzymes are not correlated. Since allosteric regulation is not effective, binding of the substrate molecules occurs mainly by spontaneous, non-activated events. The waiting times are relatively long and therefore a significant fraction ($N_0 > 150$) of the enzyme molecules is found in the waiting state, as revealed by the high peak at $\phi = 0$ in Fig. 7.2b. The active molecules are almost uniformly distributed over various cycle phases. The number of product molecules in the compartment shows only random statistical variations around a certain mean level (Fig. 7.2a, bottom); the intensity of these fluctuations agrees with the predictions of a simple Poissonian statistics for uncorrelated reaction events.

The kinetic regime of this reaction is however completely changed (Fig. 7.3) when, while keeping fixed all other reaction properties, we increase the intensity of allosteric activation, specified by the parameter ν_1. Looking at the upper part of Fig. 7.3a, one can see that the enzymes are now divided into two synchronously operating molecular groups, whose catalytic cycles are shifted by approximately half a cycle period. These groups synchronously release the product molecules and therefore the dependence of the number of product molecules in the compartment on time (Fig. 7.3a, bottom) shows a sequence of sharp spikes. The period of this sequence is twice shorter than the molecular turnover time. Note that the life-time of product molecules is short and therefore all of them die before the next spike, caused by synchronous generation by a different enzymic group, appears. The two synchronous groups are clearly seen in the histo gram of the distribution over cycle phases in Fig. 7.3b. We see also in this histogram that the number of enzymes in the waiting state is now much smaller ($N_0 < 15$), which is a consequence of strong mutual allosteric activation.

The analysis reveals[15] that the properties of the coherent reaction kinetics, realized at high intensities of allosteric regulation, significantly depend

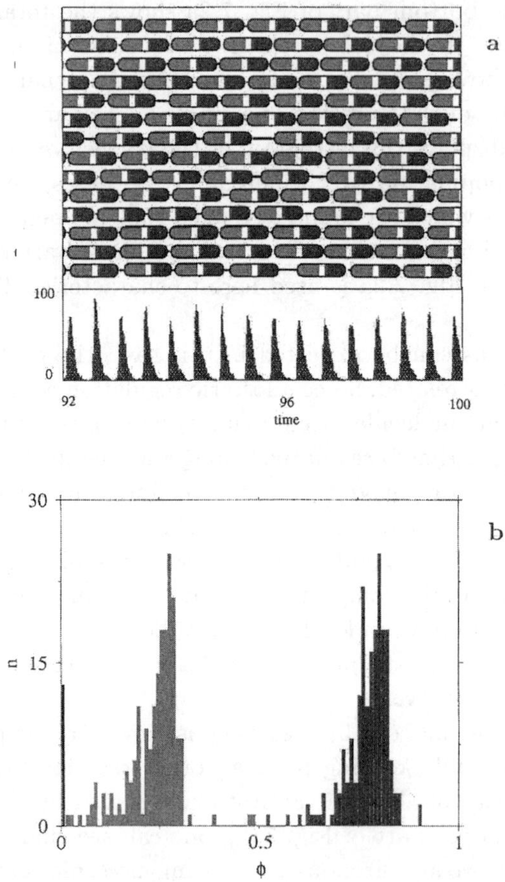

Fig. 7.3. Coherent two-group molecular dynamics of the enzymatic reaction at a higher intensity of the allosteric activation $\nu_1/\tau_0 = 0.5$. The same notations and parameter values as in Fig. 7.2. Time is measured in units of the duration τ_0 of single molecular cycle.

on the parameters of a single molecular cycle. By varying the relative time moment τ_1 inside a cycle, at which the product molecule is released, but keeping the total cycle duration $\tau_0 = \tau_1 + \tau_2$ constant, different kinetic regimes can be obtained. The behavior seen in Fig. 7.3 is realized when $\tau_1 = \tau_2 = 0.5\tau_0$ and therefore the product is released at the middle of a cycle.

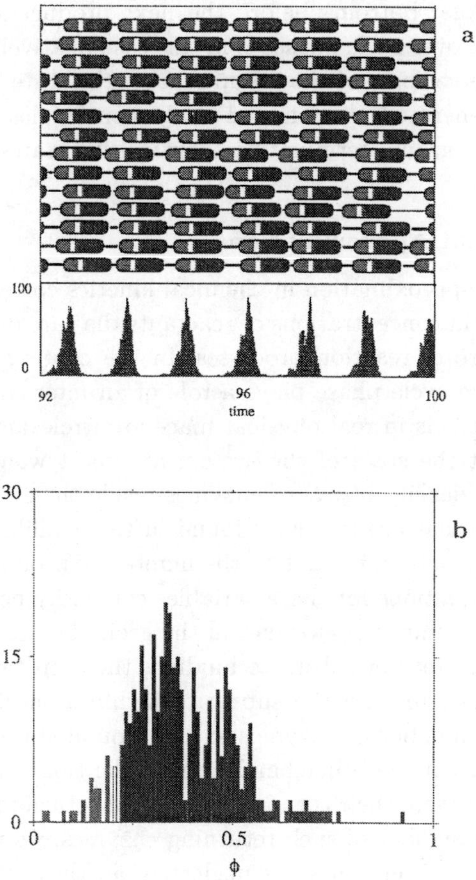

Fig. 7.4. Synchronous molecular dynamics of the entire enzymic population at $\tau_1/\tau_0 = 0.2$ and same other parameters as in Fig. 7.3. Time is measured in units of the duration τ_0 of single molecular cycle.

As an example, Fig. 7.4 shows the kinetic regime which is found for the same reaction when the products are released soon after the cycle initiation, but their release is followed by a relatively long recovery time ($\tau_1 = 0.2\tau_0$, $\tau_2 = 0.8\tau_0$). We see that now the enzymes form a single coherent group (Fig. 7.4a, top) which yields spiking in the number of product molecules at a period close to the molecular turnover time (Fig. 7.4a, bottom). Remarkably, the life-time of product molecules is still much shorter than the cycle period. Indeed, almost all product molecules which were formed in a spike still die (Fig. 7.4a, bottom) before the next product generation arrives. The distribution over cycle phases is in this case relatively wide (Fig. 7.4b). When the precursor part of the group reaches the state where the product molecules are generated, the released product molecules activate the cycles of the rest of the group, terminating their waiting states.

7.4. Mean-Field Approximation

The mean-field approximation in chemical kinetics corresponds to neglecting fluctuations in concentrations of reactants that are due to the atomistic stochastic nature of reaction processes. In the context of molecular networks, where the cycle phase plays a role of an internal dynamic variable but the distributions in real physical space are irrelevant, this approximation means that the state of the system at time t would be specified by the distribution density $p(\phi, t)$ of enzymes inside the cycle, by the number $n(t)$ of enzymes that are currently found in the equilibrium ground state, waiting to bind a substrate, and by the number of product molecules $m(t)$. The evolution equations for these variables can easily be constructed[15].

If $n(t)$ is the number of enzymes in the ground state at time t and $\nu(t)$ is the probability rate of substrate binding, then, on the average,$\nu n(t)\Delta t$ of these enzymes will bind the substrate within a short time interval Δt and, hence, the number of enzymes in the ground state will be decreased by this amount. On the other hand, during the same time interval, some enzymes will complete their turnover cycles and return to the ground state. To estimate the number of such returning enzymes, we note that, because fluctuations in turnover times are neglected, all these enzymes must have started their cycles at a delayed time moment $t-\tau_0$. Therefore, the number of such returning enzymes is equal to the number of enzymes that have bound the substrate at this time $t-\tau_0$, that is, it is given by $\nu(t-\tau_0)n(t-\tau_0)$.

If the total number of enzymes in the considered population is large and the change in the number of enzymes in the ground state within a time interval Δt is small, a continuous description can be used. The rate of change in the number of enzymes, found in the ground state, is then given by the following equation

$$\frac{dn(t)}{dt} = -\nu(t)n(t) + \nu(t-\tau_0)n(t-\tau_0). \tag{13}$$

Furthermore, the binding probability per short time step Δt is linearly dependent on the number $m(t)$ of product molecules, present at time t in the system. Therefore, for the probability rate $\nu = w/\Delta t$ at time t we get

$$\nu(t) = \nu_0 + \nu_1 m(t) \tag{14}$$

where $\nu_0 = w_0/\Delta t$ and $\nu_1 = w_1/\Delta t$.

To construct the respective evolution equation for the number $m(t)$ of product molecules, we note that on the average $\gamma m(t)\Delta t$ of these molecules will die within a short time interval Δt. On the other hand, during the same interval new product molecules will be released by enzymes. The number of such new product molecules is given by the number of enzymes that have passed within interval Δt through the state with $\phi = \tau_1$. But the latter is equal to the number of enzymes that have started their cycles at the time moment $t - \tau_1$, i.e., to $\nu(t-\tau_1)n(t-\tau_1)$. Therefore, we obtain the following equation

$$\frac{dm(t)}{dt} = -\gamma m(t) + \nu(t-\tau_1)n(t-\tau_1). \tag{15}$$

Equations (13) - (15) form a closed system. We see that in contrast to usual kinetic equations of chemical kinetics, which are also formulated in the mean-field approximation, this system of equations has some *memory* because it includes time-delayed terms. This memory is the consequence of the fact that the process is non-Markovian; that is, it is not entirely described in terms of instantaneous events, specified only by their probability rates that are independent of the process history. The intramolecular dynamics, not reducible to such rate processes, has thus been incorporated into the description.

The steady equilibrium states $n = n_0, m = m_0$ of the system are given by the fixed points of equations (13) - (15). We see that equation (13) is satisfied identically for any stationary state, whereas the two other equations give the following in the stationary case

$$\nu = \nu_0 + \nu_1 m_0, \tag{16}$$

$$-\gamma m_0 + \nu n_0 = 0. \tag{17}$$

The total number N of enzymes is equal to the sum of their number n_0 in the ground state waiting to bind a substrate, and the number N_1 of enzymes that are currently found inside the catalytic cycles. At equilibrium, N_1 can be determined from the following simple considerations: At a given time t we would find inside their cycles all those enzymes whose cycles have been initiated at the moments $t - \tau_0 < t' < t$. Under steady conditions, $\nu n_0 \Delta t$

enzymes start their cycles within each shorttime interval Δt. Therefore, the number of enzymes inside the cycles is $N_1 = \nu n_0 \tau_0$. Hence, we have

$$N = n_0 + \nu n_0 \tau_0. \tag{18}$$

Solving equations (16) - (18), we find the steady state of the system as

$$n_0 = \frac{N}{1+\overline{\nu}\tau_0}, \quad m_0 = \frac{\overline{\nu}}{\gamma} n_0, \tag{19}$$

where

$$\overline{\nu} = \frac{1}{2}\left(\nu_0 + \frac{\nu_1 N}{\gamma\tau_0} - \frac{1}{\tau_0}\right) + \sqrt{\frac{1}{4}\left(\nu_0 + \frac{\nu_1 N}{\gamma\tau_0} - \frac{1}{\tau_0}\right)^2 + \frac{\nu_0}{\tau_0}}. \tag{20}$$

This steady state may become unstable when the intensity $\nu_1$of allosteric regulation is increased.

To perform the stability analysis, we introduce small perturbations

$$m(t) = m_0 + \delta m(t), \quad n(t) = n_0 + \delta n(t). \tag{21}$$

Substituting equation (21) into equations (16) - (18) and linearizing these equations, we obtain

$$\frac{d\delta m}{dt} = -\gamma\delta m + \overline{\nu}\delta n(t-\tau_1) + \nu_1 n_0 \delta m(t - \varpi_1), \tag{22}$$

$$\frac{d\delta n}{dt} = -\overline{\nu}\delta n - \nu_1 n_0 \delta m + \overline{\nu}\delta n(t-\tau_0) + \nu_1 n_0 \delta m(t-\tau_0). \tag{23}$$

The solutions of these linear differential time-delay equations can be sought in the form $\delta m \sim e^{\lambda t}$, $\delta n \sim e^{\lambda t}$. Substituting these expressions into equations (22) - (23), we derive the characteristic equation

$$\lambda^2 + \left[\gamma + \overline{\nu}\left(1 - e^{-\lambda\tau_0}\right) - \nu_1 n_0 e^{-\lambda\tau_1}\right]\lambda + \gamma\overline{\nu}\left(1 - e^{-\lambda\tau_0}\right) = 0. \tag{24}$$

The nonpolynomial nature of the characteristic equation is typical for dynamical systems with time delays.

The roots of equation (24) are generally complex (i.e., $\lambda = \kappa + i\omega$). The steady state is stable if the real part is negative. The instability boundary is thus determined by the condition $\kappa = 0$. At this boundary, the imaginary part ω is nonzero and therefore we have a Hopf bifurcation.

Putting $\lambda = i\omega$ into equation (24), we find an equation with complex coefficients that allows one to determine the stability boundaries of the steady state in the parameter space and yields the frequency ω of the oscillations that start to grow as these boundaries are crossed. The analysis of this equation shows that it has an infinite number of solutions that correspond to the development of oscillations with frequencies lying near $\omega_k = 2\pi k/\tau_0$ with $k = 1, 2, 3, \ldots$. The oscillations give rise to the spiking regimes with different numbers k of coherent enzymic groups.

Equation (24) with $\lambda = i\omega$ has been solved numerically for different parameter values. Figure 7.5 displays the computed bifurcation diagram in the parameter plane (τ_1, ν_1). At low values of the allosteric regulation intensity ν_1, the steady state corresponding to absence of oscillations is stable. If we increase the parameter ν_1, going along the vertical direction in the bifurcation diagram, one of the instability boundaries would be crossed. Above this boundary, the steady state is unstable with respect to development of periodic oscillations.

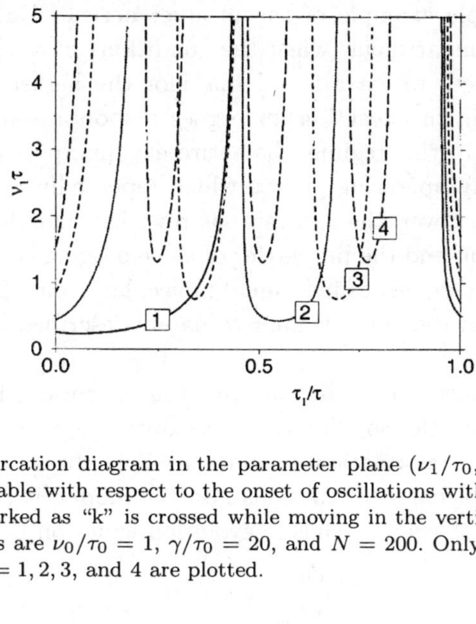

Fig. 7.5. The bifurcation diagram in the parameter plane $(\nu_1/\tau_0,\ \tau_1/\tau_0)$. The steady state becomes unstable with respect to the onset of oscillations with k coherent groups, when the curve marked as "k" is crossed while moving in the vertical direction. Other reaction parameters are $\nu_0/\tau_0 = 1$, $\gamma/\tau_0 = 20$, and $N = 200$. Only the first instability boundaries with $k = 1, 2, 3$, and 4 are plotted.

Above the first crossed boundary, the system is already found not in the steady, but in the oscillatory state. Hence, other instability boundaries of the steady state, that lie above the first boundary, are not relevant. Different curves in Figure 7.5 correspond to the onset of oscillations with different basic frequencies $\omega_k = 2\pi k/\tau_0$; that is, with different numbers k of coherent enzymic groups. The number of such groups in the first unstable mode is determined by the parameter τ_1. Inside the intervals $0 \le \tau_1/\tau_0 \le 0.34$ and $0.955 \le \tau_1/\tau_0 \le 1$ the first crossed boundary corresponds to $k = 1$ and hence synchronous oscillations of the entire enzymic population are expected. Inside the intervals $0.34 \leq \tau_1/\tau_0 \leq 40$ and $0.65 \leq \tau_1/\tau_0 \leq 0.78$, spiking with three coherent groups ($k = 3$) is expected, etc.

The behavior of the system above the instability boundary has been studied by numerical integration of full nonlinear delay equations of the mean-field approximation[15]. This investigation has shown that the considered instability indeed leads to the development of persistent oscillations with the frequencies and numbers of enzymic groups determined by the linear stability analysis just presented. However, the bifurcation may be subcritical, so that hysteresis is observed.

7.5. Further Theoretical Developments

Analyzing a simple example of the product activated allosteric reaction (8) in a small compartment, when the conditions of a molecular-network regime are satisfied, we have found that, for the higher intensities of allosteric regulation, this reaction undergoes a spontaneous transition to a coherent regime. In this regime, the entire enzymic population forms several synchronously operating groups which repeatedly activate the cycles of each other. Moreover, the analysis has revealed that the exact structuring of a population and the properties of the emerging coherent regime are determined by the details of individual molecular cycles. The transition to coherent molecular dynamics is approximately described by the mean-field approximation.

We have assumed above that intramolecular motions inside a turnover cycle are deterministic, so that the cycle duration τ_0 is fixed. In reality, motion along the conformational coordinate ϕ is always stochastic. To incorporate intramolecular fluctuations into the model, it can be assumed[15] that the phase ϕ obeys a stochastic evolution equation

$$\frac{d\phi}{dt} = v + \eta(t), \tag{25}$$

where $\eta(t)$ is a white noise of some given intensity.The product molecule is released when the phase state ϕ_1 is passed and the cycle is ended when the phase $\phi = 1$ is reached. The cycle time τ is not fixed and fluctuates with some statistical dispersion $\xi = (1/\tau_0)\sqrt{< \tau^2 > - \tau_0^2}$ around its mean value $\tau_0 =< \tau >= 1/v$.

Numerical investigations have shown[15] that relatively weak intramolecular fluctuations do not destroy a transition to coherent molecular dynamics. The synchronization can be observed even when the statistical dispersion ξ of turnover times is about 20%.

Another simplification in the above model was that related to the description of allosteric activation in the system. For simplicity, it was assumed here that, when m regulatory product molecules are present in the volume, this increases the probability of cycle initiation according to equation (10). The actual mechanism of allosteric activation is more complex. A regulatory molecule binds to an enzyme at some special binding site. This leads to a conformational change in the enzyme molecule and an increase in the probability of substrate binding and, thus, of cycle initiation. A more detailed model explicitly includes binding of regulatory (product) molecules and the process of their spontaneous detachment[15]. Note that here some further details of the operation of molecular machines are essential. Can the regulatory molecule bind to an enzyme only when it is in its ground state, waiting to bind a substrate? Or it can bind to an enzyme even when it is inside its catalytic turnover cycle? Does the regulatory molecule immediately detach from the enzyme when the cycle is started or it can continue to sit on the enzyme and remain there even the next cycle is initiated? Such questions, which are not even asked in the standard approach to enzymatic kinetics, become important. Depending on these mechanistic details, the synchronization transition can be facilitated or may get more difficult[15].

Not only allosteric activation, but also allosteric inhibition is possible. In some enzymatic reactions, the product is inhibiting the reaction, either by directly blocking the binding site for the substrate or by producing a conformational change that makes substrate binding less probable. Investigations of models with allosteric product inhibition have shown that a transition to coherent molecular dynamics can be observed in such systems[16]. The difference is that, under product inhibition, only synchronous spiking of the entire population (with just one coherent enzymic group) can take place.

Many enzymes consist of several functional subunits. Each subunit is effectively an independent enzyme that can bind a substrate and perform its own catalytic turnover cycle. Cooperative interactions between functional subunits in an enzyme are usually found. When the cycle of one subunit is initiated, this either enhances the probability of cycle initiation in other subunits in the same molecule or surpresses such cycles. The models with several interacting functional subunits and allosteric product activation or inhibition have been investigated[17]. An interesting aspect of such models is that synchronization of turnover cycles between subunits inside the same enzyme molecule is possible. Sometimes such intramolecular synchronization can take place under the conditions when allosteric product activation favors formation of several enzymic groups. When this occurs, the tendency towards internal synchronization competes with the processes preferring phase difference in the operation of functional enzymic subunits[17].

Finally, it was also found that allosteric regulation is not necessary to produce molecular synchronization. Investigation of the reaction schemes where allosteric regulation was absent, but a fraction of the product was converted back into the substrate for the same enzymic reaction, have shown that coherently operating enzymic groups can spontaneously develop in such systems[18,19].

The kinetic model, treating an individual enzyme molecule as a cyclic automaton, is only a simple approximation. Experiments with single enzymes yield information that the operation of an enzyme can be more complicated[20]. The experimental data for some enzymes could be interpreted assuming that an enzyme molecule possesses two catalytic turnover cycles of different duration, and the shorter cycle can be started soon after the product release, when the the enzyme remains in an "active" state[21].

7.6. Coherence in Cross-Coupled Dynamical Networks

So far, only relatively simple reaction schemes with a single kind of enzymes and a single kind of regulatory molecules were considered. The actual reaction networks of a living cell are much more complicated. They include many different kinds of enzymes and regulatory molecules. The same regulatory molecules can play an activatory role for some of the enzymes and inhibit the activity of the others.The system can thus be viewed as a supernetwork, formed by a large number of different interacting molecular networks.The question is whether synchronization and just-in-time production can also spontaneously develop in such complex systems.

Although the studies of synchronization in complex biochemical networks have not yet been performed, some insight into the synchronization behavior of cross-coupled networks can be obtained from the analysis of a simple model describing colelctive dynamics in an ensemble of cross-coupled networks formed by idealized neurons[22]. Here, a neuron is an automaton that can change its activity in response to signals arriving from other neurons. A network consists of a set of such elements linked through activatory or inhibitory connections. The dynamics of a network with arbitrary asymmetric connections is typically characterized by an irregular sequence of complex activity patterns.

The collective dynamics of the ensemble is described by the following algorithm. Each of N identical networks in the ensemble consists of K

neurons. At the next discrete time moment $t+1$, the activity x_k^i of a neuron $k = 1...K$ belonging to a network $i = 1...N$ is

$$x_k^i(t+1) = (1-\varepsilon)\theta(h_k^i) + \varepsilon\theta\left(\sum_{j=1}^{N} h_k^j\right) \tag{26}$$

where

$$h_k^i = \sum_{l=1}^{K} J_{kl} x_k^l(t) \tag{27}$$

is the signal arriving to this neuron from all other elements of the same network, J_{kl} are the connection weights (the same for all networks), and $\theta(z)$ is a sigmoidal function, such as $\theta(z) = 0$ for $z < 0$ and $\theta(z) = 1$ for $z \geq 0$.

The two terms in the right side of equation (26) have a clear interpretation. The first of them represents the individual response of a neuron to the total signal received from all other elements in its own network. The second term depends on the global signal obtained by summation of individual signals received by neurons occupying the same positions in all networks of the ensemble. The parameter ε specifies the strength of global coupling. When global coupling is absent ($\varepsilon = 0$), the networks forming the ensemble are independent. On the other hand, at $\varepsilon = 1$ the first term vanishes and the states of respective neurons in all networks must be identical since they are completely determined by the same global signal. For $0 < \varepsilon < 1$, the ensemble dynamics is governed by an interplay between local coupling inside the networks and global coupling between them.

This model exhibits, under increasing the global coupling intensity, a spontaneous transition to a coherent collective behavior[22]. The transition is characterized by the formation of coherent network clusters followed by complete synchronization of all networks in the ensemble. It takes place already at low intensities of global coupling and is observed under an arbitrary choice of the connection weights in elementary networks and for random initial conditions. In numerical simulations, 100 networks each consisting of 50 neurons were taken[22].

Figure 7.6 shows typical results of the simulations. The integral time-dependent activity $u_i(t) = \sum_{k=1}^{K} x_k^i(t)$ of 10 selected networks in the ensemble is displayed here (this presentation is reminiscent of the EEG records where the signal coming from each electrode is an average of the neural activity in the respective brain area). The dynamical patterns of network activity are always very irregular. However, depending on the intensity of coupling between the networks, a varying degree of coherence can be discerned in these patterns.

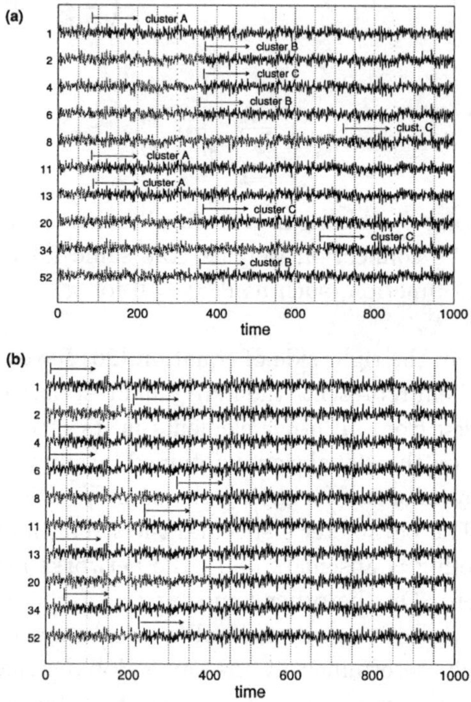

Fig. 7.6. Time-dependent integral activity $u_i(t)$ of 10 selected networks in an ensemble of 100 identical networks each consisting of 50 neurons for different intensities of global coupling, corresponding to (a) dynamical clustering ($\varepsilon = 0.35$) and (b) complete synchronization ($\varepsilon = 0.5$). Synchronization of each signal begins at the corresponding. In part a, clusters are identified by random initial conditions, random choice of synaptic connection weights inside a network.

At a low coupling intensity, the activity of different networks is not coherent. When, however, a larger intensity of coupling is chosen, the behaviour of the ensemble is qualitatively changed. As time goes on, the activity signals generated by some of the networks suddenly become exactly identical. Eventually, the whole ensemble breaks down into several coherent clusters, so that for all networks belonging to the same cluster the total activity is the same at all times (Fig. 7.6a). At higher intensities of global coupling (Fig. 7.6b), the activity of all networks in the ensemble becomes coherent. Remarkably, the coherent signals are still very complex and apparently chaotic.

The robustness of synchronization in coupled neural networks, which has been observed under an arbitrary choice of activatory and inhibitory connections between neurons and for the networks and ensembles of different sizes, may serve as an indication that the coherent dynamics is typical for the networks of various origins.

7.7. Discussion

In the introduction, a living cell has been compared with a large industrial factory where thousands of interconnected assembly lines are running in parallel to make a great variety of products. Some of these assembly lines may operate in the synchronous mode, while the operation of others can be asynchronous. The employment of synchronous network organization (or just-in-time production) has several important advantages. It can lead to a significant increase in the production efficiency of a reaction, as compared with its output in the usual noncorrelated kinetic regime. Moreover, the coherent molecular networks are much more flexible in the functional sense.

Already the analysis of a simple product-activated enzymatic reaction shows that the same reaction can exhibit different forms of dynamic organization depending on the parameters of individual molecular cycles. The entire enzymic population can break into several synchronously operating groups which activate one another. These groups build, on the basis of the same underlying molecular network, various functional networks where each node is formed by a group of coherently operating elements and the links between the nodes indicate interactions between different functional groups. This phenomenon of emergent functional self-organization of a molecular network shall become even more complex and significant when the networks consisting of several kinds of enzymes and comprising both activatory and inhibitory interactions are considered.

The networks operating in a coherent dynamical mode generate periodic frequent spikes of product molecules. The presence of such frequent generators inside a cell may be important because they can be used to externally synchronize other reactions, if the product molecules of a coherently operating network are used as substrate or regulatory molecules for these other reactions. Switching of the generation frequency can then control a transition to a different organization mode of the larger biochemical system.

Furthermore, not only the frequency but also the phase difference are important for controlling the coherent molecular dynamics. Indeed, when a short intensive spike of regulatory molecules arrive, it can execute its regulatory action only if the respective enzyme molecules are found at this time in a stage of their molecular cycles where they are responsive to a regulation. For the complicated branched reaction schemes it might mean that the same reaction would yield different kinds of final products depending on the fine phase tuning, if this reaction is operating in the non-classical regime of a molecular network.

References

1. P. Stange, D. H. Zanette, A. S. Mikhailov, B. Hess, Biophys. Chem. 72 (1998) 73.
2. A. Goldbeter, Biochemical Oscillations and Cellular Rhythms, Cambridge University Press, Cambridge, 1996.
3. B. Hess, Quart. Rev. Biophys. 30 (2) (1997) 121.
4. B. Hess, A.S. Mikhailov, Ber. Bunsenges. Phys. Chem. 98 (1994) 1198
5. B. Hess, A.S. Mikhailov, J. Theor. Biol. 176 (1995) 181.
6. H. Gutfreund, Kinetics for the Life sciences: Receptors, Transmitters, and Catalysts (Cambridge Univ. Press, Cambridge, 1995)
7. M. von Smoluchowski, Phys. Z. 17 (1916) 557.
8. M. Eigen, Z. Phys. Chem. (München) 1 (1954) 176.
9. J.B. Wittenberg, B.A. Wittenberg, Annu. Rev. Biophys. Chem. 19 (1990) 217.
10. L. Stryer, Biochemistry, 3rd edn. (Freeman, New York, 1988).
11. L.A. Blumenfeld, A.N. Tikhonov, Biophysical Thermodynamics of Intracellular Processes. Molecular Machines of the Living Cell (Springer, Berlin, 1994).
12. B. Hess, A.S. Mikhailov, Science 264 (1994) 223
13. B. Hess, A.S. Mikhailov, Biophys. Chem. 58 (1996) 365
14. A.S. Mikhailov, B. Hess, J. Phys. Chem. 100 (1996) 19059.
15. P. Stange, A. S. Mikhailov, B. Hess, J. Phys. Chem. B 102 (1998) 6273.
16. P. Stange, A. S. Mikhailov, B. Hess, J. Phys. Chem. B 103 (1999) 6111.
17. H.-P. Lerch, P. Stange, A. S. Mikhailov, B. Hess, J. Phys. Chem. 106 (2002) 3237
18. P. Stange, A. S. Mikhailov, B. Hess, J. Phys. Chem. B 104 (2000) 1844.
19. A. S. Mikhailov, P. Stange, B. Hess, in Single Molecule Spectroscopy: Nobel Conference Lectures, eds. R. Rigler, M. Orrit, T. Basche (Springer, Berlin, 2001) p. 277.
20. L. Edman, R. Rigler, Proc. Natl. Acad. Sci. USA 97 (2000) 8266.
21. H.-P. Lerch, A. S. Mikhailov, B. Hess, Proc. Natl. Acad. Sci. USA 99 (2002) 15410.
22. D. H. Zanette, A. S. Mikhailov, Phys. Rev. E 58 (1998) 872.

CHAPTER 8

CELL IS NOISY

Tatsuo Shibata
Department of Mathematical and Life Sciences, University of Hiroshima,
1-3-1, Kagamiyama, Higashi-Hiroshima, 739-8526, Japan
E-mail: shibata@hiroshima-u.ac.jp

The chemical reaction is inherently stochastic. Such stochasticity is particularly prominent in small systems, such as biological cells. In this chapter, we show how gene expression and signal transduction reactions are noisy based on the recent theoretical and experimental studies.

8.1. Introduction

The cell is a small system. Although the number of chemical species in a cell is quite large, the number of molecules of each type is often small[1]. For instance, the number of mRNA molecules of a given type is typically fewer than ten copies per a cell. In procaryote cells, the number of a particular kind of regulatory protein, which controls the expression of a gene, is typically much less than one hundred. Time evolutions of chemical reaction processes are considered as stochastic processes. Therefore, under these conditions, strong fluctuations in the number of molecules must be present inside cells. We call this "molecular noise". For some systems, large fluctuations in the numbers of a protein may be dangerous. The cellular systems should have reliability in order to properly respond to external stimuli. How does a cell achieve such reliability despite large fluctuations? The cell systems should possess some mechanisms which would allow them to reduce molecular noise. In some cases, large fluctuations may be advantageous to perform the functions.

Recently, time-resolved experimental measurements of gene expression and signal transduction systems have been performed. The existence of strong noise in gene expression has been demonstrated experimentally[2,3,4,5].

The cellular signal transduction systems are also inherently noisy, which has been shown experimentally by single-molecule and single-cell analysis[6,7,8,9]. For gene expression systems, the noise of expression has been studied theoretically[10,11,12], suggesting the noise can be much larger than the Poissonian noise. It has been pointed out that the two distinct noise sources contribute to the noise of each component: one is the noise inherent in its own reaction (intrinsic noise), and the other one is the noise generated in other chemical components that affects the reaction (extrinsic noise)[4,13,14]. For signal transduction systems, Oosawa discussed that the intracellular noise is hierarchically organized from thermal fluctuations to spike-like large fluctuations, which produce spontaneous signals to change the behavior of swimming cells such as bacteria and paramecia[15,16]. Recently, the noise in signal transduction was discussed suggesting the large amplification results in the generation of strong random fluctuations in the output signal[17,18]. One essential feature of biological signal transduction systems is to amplify small changes in input signals[19]. It has been shown theoretically that how the abrupt response of ultrasensitive signal transduction reactions results in both generation of large inherent noise and high amplification of input noise[20]. This relation between the response and the fluctuation is a variant of fluctuation-dissipation theorem in nonequilibrium statistical physics[14,20]. The relation between fluctuation and evolutional responses has also been studied[21].

In this chapter, the stochastic properties of gene expression and signal transduction reactions are reviewed. In Section 8.2, the origin of molecular noise is briefly reviewed and the method to simulate Poisson stochastic processes is shown. In Section 8.3, the model of gene expression to study the stochastic behavior is shown. Then, it is shown that the noise in gene expression is larger than the Poisson noise, and the autoregulation reduces the noise. In Section 8.4, the noise generated by a typical signal transduction reactions is shown. In particular, the gain-intrinsic noise relation is indicated. In Section 8.5, the propagation of noise in a reaction network is studied. Between the propagation of noise and the gain of signal, the gain-extrinsic noise relation is show to be hold.

8.2. Origin of Molecular Noise

The number of given type of molecule changes by chemical reactions. According to the theory of chemical reactions, when two types of molecules collide, if these two molecules have enough energy to across a transition state, the molecules change from the state of reactants to the state of products so that the reaction takes place. If the transition of the two molecules is fast enough, these processes can be described by a deterministic dynamical system, employing Newtonian mechanics. In cells, each molecule can be quite huge, such as enzymes or protein machinery. Thus, the dynamical system consists of a huge number of variables describing the motion of the atoms. Moreover, to describe the evolution of the number of a molecule, it is necessary to have knowledge of the time when a collision take place and how much energy the two molecules have when they collide. In a cell, a huge number of chemical species are densely packed. The motion of those molecules could determine the time of the collision and the energy of the two molecules. In practice, however, it is impossible to describe and predict the evolution of the number of a given type of molecule according to such a microscopic description.

Most of the processes mentioned above are taking place much faster than the change in the number of a molecule, and a huge number of molecules are involving those processes. Thus, within the time interval of occurring one reaction of a given type, a huge number of events have already taken place in a cell, that make us possible to adopt statistical descriptions. If the information about a particular reaction disappear fast enough due to the huge number of events, the two successive reactions of a given type can be considered as *statistically independent.* This means that the probability of occurring two successive reactions at particular times t_1 and t_2 is given just by the product of the two probabilities of occurring the reaction at time t_1 and occurring the reaction at time t_2. The time interval of successive reactions obeys a exponential distribution. In such a process, the evolution in the number of a molecule can be described by *a Poisson process.*

In a Poisson process, when the probability of having a reaction within a second is r, the statistical variation of the number of reaction taking place within the same time interval, measured by the standard deviation, is $\sqrt{r}$. Therefore, the relative strength of the statistical variation compared with the mean number of reaction within a second is given by $1/\sqrt{r}$, which decreases as the reaction probability increases. If the probability is large enough, one can neglect the statistical variation in the number of reactions occurring per a second. Then, the evolution of the number is again described by a deterministic dynamical system, such as ordinary differential equations. Such descriptions are expected when the number of molecules of a given type is large.

Let us consider a reaction taking place between molecules A and B producing C; A + B → C. The probability of having a reaction per a second in a cell is given by

$$Vk\frac{N_\mathsf{A}}{V}\frac{N_\mathsf{B}}{V} \tag{1}$$

where k is the probability of reaction taking place between particular two molecules of A and B, V is the volume of the cell, N_A and N_B is the number of each molecule at a given time[22]. If N_A and N_B are large enough, the statistical variation of the number of reactions per a second can be neglected. Then, the number of reaction occurring per a second per a unit volume can be expressed by

$$kx_\mathsf{A}x_\mathsf{B} \tag{2}$$

where $x_\mathsf{A} = \frac{N_\mathsf{A}}{V}$ and $x_\mathsf{B} = \frac{N_\mathsf{B}}{V}$. Thus, k can be considered as the rate constant of the reaction. So far the value of k is accessible only as a rate constant by macroscopic experiments. In the stochastic descriptions, experimentally obtained rate constants are interpreted as the probability of having the reaction per a second amongst particular molecules.

In a Poisson process, the time interval between succeeding two evens follows an exponential distribution. Thus, Poisson processes are numerically simulated by generating a series of random numbers that satisfy the exponential distribution. Following this idea, the Gillespie's algorithm gives a numerical scheme to simulate time evolutions of stochastic chemical reactions[23].

8.3. Stochastic Gene Expression

In this section, we study a single gene expression and a gene expression with autoregulation. In order to study these systems, we introduce models and equations (stochastic kinetic equation) that describe the systems. Then, the average gene expression and its molecular noise are presented.

The evolution of the number of a given chemical species due to chemical reactions can be described by Markov processes, which can be displayed, for instance, by the chemical master equation[22]. When the number of molecule is not extremely small, it is expected that the evolution can be described by stochastic kinetic equations, or chemical Langevin equations, which are stochastic differential equations[22,24]. In chemical Langevin equations, the number of molecules is described by continuous real numbers rather than discrete integer numbers.

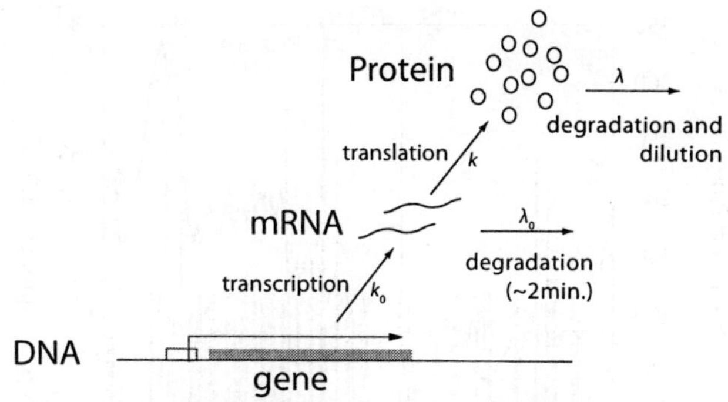

Fig. 8.1. A single gene expression in procaryote. The important parameters are the transcription rate k_0, the translation efficiency b defined by the mean protein product per single mRNA, and the degradation rate of the protein products λ. The translation efficiency b is the translation rate k divided by the degradation rate λ_0 of mRNA.

The models that include detailed description of gene expression processes may reproduce the behavior in detail. However, such detailed models may be difficult to be applied to large scale gene regulatory networks. Here, instead of studying detailed model, we show a model that describes gene expression processes as simple as possible, but yet it captures essential characteristics of their stochastic nature.

8.3.1. *Noise in Single Gene Expression*

We start with a system of a gene expression without any regulations (Fig.8.1). In Fig.8.2, the time evolution of the numbers of protein and mRNA products calculated numerically is shown. The parameters are chosen so that average number of protein product is one hundred. However, the number fluctuates around the mean number quite strongly. In order to study these system theoretically, several models have been proposed[11,12,13]. Even this simplest case, a lot of processes are included, such as binding of RNA Polymerase (RNAP), the initiation of transcription, the transcription progression of RNAP, the binding of ribosome and RNase on mRNA, and transcription progression of ribosomes[11]. We begin with the model proposed by Thattai and vanOudenarrden[12], in which the state of the gene expression is described by the number of mRNA and protein products. Another processes included in the gene expression could affect the behavior, but its contributions especially on the noise of gene expression are considered to be minor.

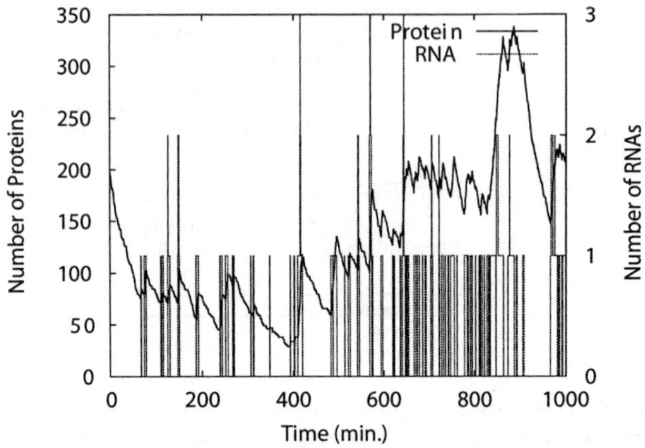

Fig. 8.2. Stochastic time evolution of the number of proteins and RNA transcripts. The ensemble average number of proteins is 100.

In the case of procaryote, the time scale of mRNA degradation is about a few minutes whereas the time scale of protein degradation is about one hour. Thus, although the number of mRNA could be a few within a cell and this suggest strong molecular noise of mRNA, the concentration of mRNA can be eliminated from the description of the gene expression. Then, the evolution of the protein product can be described by a single variable stochastic kinetic equation. Instead of explicitly considering the concentration of mRNA, we consider the production rate of the protein fluctuates in time. For the mathematical detail of this approximation, see Refs. 25 and 26. The production and the depletion of protein product X of the gene expression is described by the following reaction scheme:

$$\mathsf{G_X} \overset{k_0 b + \sqrt{2D}\eta(t)}{\longrightarrow} \mathsf{X}, \quad \mathsf{X} \overset{\lambda x}{\longrightarrow} \phi \tag{3}$$

with

$$D = k_0 b^2 \tag{4}$$

where x is the number of protein X products, k_0 is the transcription rate, b is the translation efficiency that is the mean number of protein product per a single transcript, and λ is the degradation rate of the protein. Here, $k_0 b + \sqrt{2D}\eta(t)$ is the fluctuating reaction rate[25] that gives the probability of synthesizing a molecule of protein X per unit time. $\eta(t)$ is the Gaussian white noise with $\langle \eta(t)\eta(t')\rangle = \delta(t - t')$. The constant $D = k_0 b^2$ gives the strength of the fluctuation of the reaction rate. The noise term in the

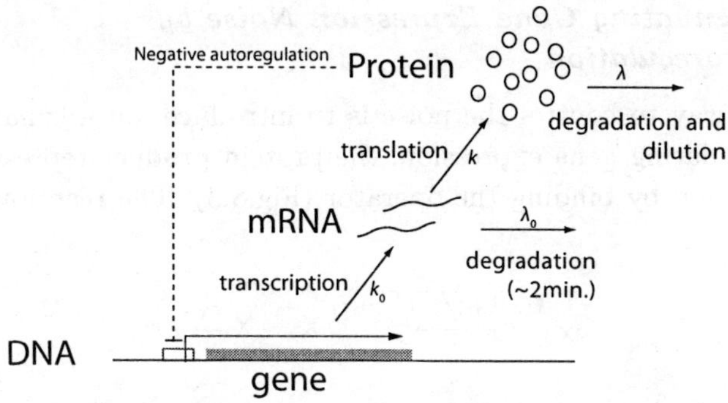

Fig. 8.3. A gene expression with autoregulation.

fluctuating reaction rate yields a contribution from the fluctuation in the number of mRNA.

The evolution of the number of protein product X is described by a chemical master equation or a chemical Langevin equation. Let x be the number of the protein product X in a cell. The chemical Langevin equation describing the evolution of x is written by

$$\dot{x} = k_0 b - \lambda x + \sqrt{k_0 b(1+2b) + \lambda x} \cdot \xi(t) \tag{5}$$

Here, $\xi(t)$ is the Gaussian white random noise with $\langle \xi(t)\xi(t')\rangle = \delta(t-t')$.

From this expression, the mean value and the variance can be calculated at

$$m = \langle x \rangle = \frac{k_0 b}{\lambda} \tag{6}$$

and

$$\sigma^2 = \langle x^2 \rangle - \langle x \rangle^2 = (1+b)\langle x \rangle. \tag{7}$$

The same expression can be obtained by solving the master equation[12], under the condition that the degradation rate of protein is much smaller than mRNA. Notice that in the case of the Poisson distribution, the variance is equal to the average, i.e., $\sigma^2 = \langle x \rangle$. Therefore, the molecular noise of gene expression is stronger than the Poisson distribution. Eq. (7) is verified experimentally by van Oudenaarden, et al.[3].

8.3.2. *Attenuating Gene Expression Noise by Autoregulation*

One of the way to control the noise is to introduce autoregulation[2,12]. In the autoregulating gene expression, the protein product represses its own gene expression by binding the operator (Fig.8.3). The reaction scheme is given by

$$\mathsf{G_X} \overset{k(x)b+\sqrt{2D(x)}\eta(t)}{\longrightarrow} \mathsf{X}, \quad \mathsf{X} \overset{\lambda x}{\longrightarrow} \phi \tag{8}$$

with

$$k(x) = k_0/(1 + (x/\phi)^p) \quad \text{and} \quad D = k(x)b^2 \tag{9}$$

where x is the number of X molecules, $k(x)$ gives the transcription rate for a given number of x. Here, k_0 is a constant number, ϕ is the dissociation constant of the protein to the operator, and p is the Hill coefficient which specifies the sensitivity to the change of the number x, respectively. As before, b is the translation efficiency that is the mean number of protein product per a single transcript, and λ is the degradation rate of the protein. $\eta(t)$ is the Gaussian white noise with $\langle \eta(t)\eta(t')\rangle = \delta(t-t')$.

The transcription rate $k(x)$ is obtained by considering the process of the regulatory proteins binding to and unbinding from operator explicitly[27,26]. This process also contributes to the noise of gene expression. However, if the binding and unbinding is occurring frequently enough, the noise can be negligible. (the noise strength compared with the mean value is proportional to $\sqrt{k_d}/k_a$, in which k_d is the unbinding rate and k_a is the binding rate.)

The evolution of the number x of the protein product is described by the equation

$$\dot{x} = \frac{k_0 b}{1 + (x/\phi)^p} - \lambda x + \sqrt{\frac{k_0 b}{1 + (x/\phi)^p}(1 + 2b) + \lambda x} \cdot \xi(t) \tag{10}$$

where $\xi(t)$ is the Gaussian white random noise with $\langle \xi(t)\xi(t')\rangle = \delta(t-t')$.

In order to study the characteristics of the autoregulation, consider the case when the dissociation constant is small, i.e., $\phi \ll 1$. Applying the linear noise approximation method to Eq.(10), the variance can be obtained. First, the average number $\langle x\rangle$ is obtained from the macroscopic equation:

$$\langle x\rangle = \left(\frac{k_0 b \phi^p}{\lambda}\right)^{\frac{1}{1+p}}. \tag{11}$$

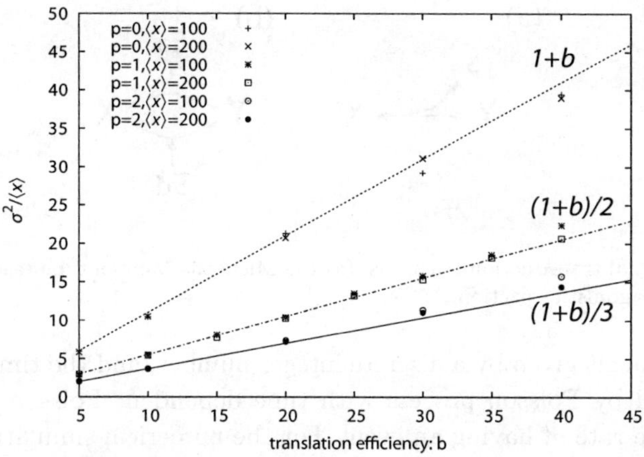

Fig. 8.4. The ratio between the variance and the average number of protein products plotted as a function of the translation efficiency b. As the Hill coefficient p of the autoregulation increases from $p = 0$ to 2, the noise intensity is suppressed.

Linearizing Eq.(10) around the average number $\langle x \rangle$, and substituting the average number $\langle x \rangle$ into x in the third term in Eq.(10), it follows that

$$\dot{\delta x} = -\lambda(1+p)\delta x + \sqrt{2\left(\frac{k_0 b \phi^p}{\lambda}\right)^{1/(1+p)} \lambda(1+b)} \cdot \xi(t), \qquad (12)$$

where $\delta x = x - \langle x \rangle$. Solving this equation, the variance is obtained as

$$\sigma^2 = \langle x^2 \rangle - \langle x \rangle^2 = \frac{1+b}{1+p}\langle x \rangle \qquad (13)$$

This first order approximation indicates that as the Hill coefficient p increases, the noise is reduced for the same value of the average number. Note that this autoregulatory system is not considered as a kind of negative feed-back system, since there is no mechanism to reduce the deviation from a particular reference value.

In Fig.8.4, the variance and the average value is calculated numerically and the ratio between them is plotted as a function of the translation efficiency b. For the numerical study, we adopt the following reaction scheme,

$$\mathsf{G_X} \xrightarrow{k(x)} \mathsf{G_X} + \mathsf{M_X}, \quad \mathsf{M_X} \xrightarrow{\lambda_\mathsf{M}[\mathsf{M_X}]} \phi \qquad (14)$$

$$\mathsf{M_X} \xrightarrow{b\lambda_\mathsf{M}} \mathsf{M_X} + \mathsf{X}, \quad \mathsf{X} \xrightarrow{\lambda x} \phi \qquad (15)$$

where λ_M is the degradation rate for mRNA transcript, and $[\mathsf{M_X}]$ is the number of mRNA transcript. In the numerical calculation the number of

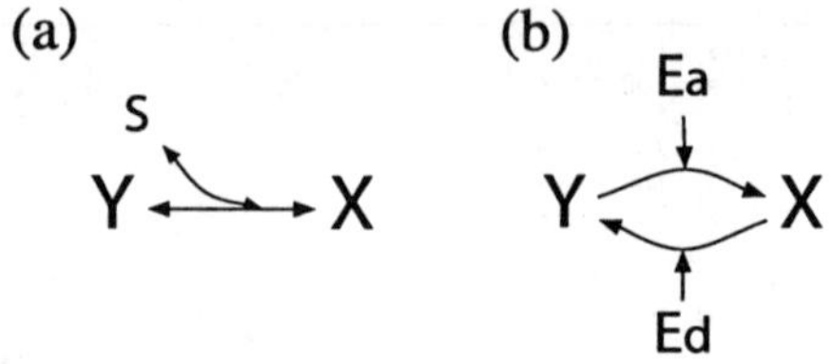

Fig. 8.5. Signal transduction reactions: (a) the Michaelis-Menten type reaction, (b) the push-pull antagonistic reaction.

each molecule is given by a discrete integer number, and the time evolution is described by Poisson process with time dependent Poisson parameter, which is the rate of having an event. For the numerical simulation of Poisson process, the Gillespie's numerical algorithm was used[23]. The numerical results exhibit good agreement with the theoretical result based on the chemical Langevin equation.

8.4. Noisy Signal Amplification in Signal Transduction Reactions

Many cellular processes respond quickly to internal and external variations by using chemical reaction networks. For instance, chemotactic amoeba cells such as *Dictyostelium discoideum* responds to move up a shallow chemoattractant gradient within a minutes. This time scale is much faster than the time scale of gene expression responses, which is slower than a tens of minutes. Signal transduction networks are responsible for these quick responses. These networks are typically consists of interactions between proteins.

The chemotactic amoeba cells can move up shallow chemoattractant gradients, which is typically 2-5 % of the cell length[7]. It indicates that they can compare and process extremely small differences in the concentrations of extracellular stimuli. One strategy to respond to such small signals is to adopt switch-like reactions, which can generate abrupt responses from a small change in the input stimuli. There are several reactions that sharpen the responses, such as cooperative reactions in single proteins[19,28,29], and push-pull antagonistic reaction[30,31]. In these reactions, the response is switch-like, with a threshold in the concentration of stimuli. In a cascade of such switch-like reactions, as observed in mitogen activated protein kinase cascade, the amplification of the whole cascade can be much larger[32]. Is it more appropriate for the cellular systems to have reactions that exhibit steep responses in order to generate all-or-none cellular behaviors?

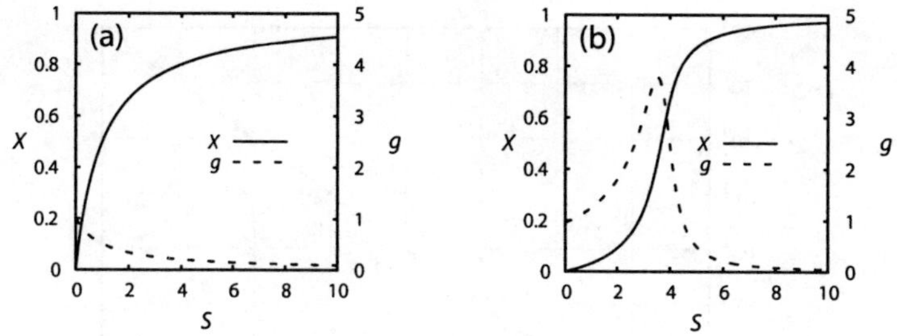

Fig. 8.6. Ultrasensitive responses in signal transduction reactions. The fractional concentration of the output signal X (left axis) and the gain g (right axis) are plotted as functions of the concentration of signal molecule S, (a) the Michaelis-Menten type reaction, (b) the push-pull antagonistic reaction.

As we have seen in the gene expression processes, cellular processes are inherently noisy. Such noisy characteristic of reactions may affect the behaviors of switch-like signal transduction reactions. Therefore, here we study how the noise generated a reaction relates to the amplification of signal[20].

As examples, we study two types of reactions which is always found in signal transduction cascades: one is the Michaelis-Menten-type reactions, and the other one is a reaction in which a messenger is activated and deactivated cyclically by a pair of opposing enzyme, as observed in a combination of kinase and phosphatase reactions.

The Michaelis-Menten type reaction is the simplest signal transduction reaction that behaves as a molecular switch:

$$\mathrm{S} + \mathrm{Y} \leftrightarrow \mathrm{X} \tag{16}$$

where S is the signaling molecule that binds to the inactive state Y so that the protein is switched on to the active state X (Fig.8.5(a)). This best-known reaction gives rise to the Michaelis-Menten kinetics (Fig.8.6(a)).

The push-pull antagonistic reaction is the simplest example of cyclic modification reactions that can show a sharp response[30,31]. In the push-pull reaction, the signaling molecule, that is an enzyme, switches its substrate protein from an inactive state to an active state, whereas another enzyme switches the protein off (Fig.8.5(b)). Each step is characterized by Michaelis-Menten kinetics:

$$\begin{array}{c} \mathrm{Y} + \mathrm{E_a} \leftrightarrow \mathrm{YE_a} \rightarrow \mathrm{X} + \mathrm{E_a} \\ \mathrm{X} + \mathrm{E_d} \leftrightarrow \mathrm{XE_d} \rightarrow \mathrm{Y} + \mathrm{E_d} \end{array} \tag{17}$$

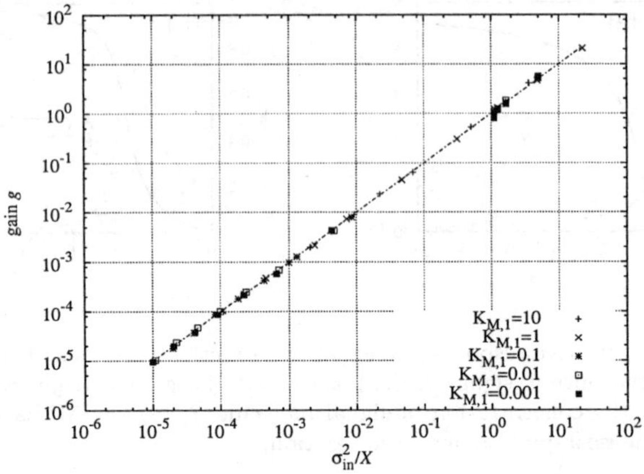

Fig. 8.7. The gain-intrinsic noise relation of signal transduction systems. The gain g is plotted as a function of the variance divided by the mean number. Changing the concentration of the input signal S, the gain, the variance and the mean number are obtained by performing stochastic simulations of schemes, as shown in Eq.(17) according to the Gillespie's numerical algorithm [23]. The Push-pull reaction: The Michaelis constant is $K_{\rm a}$ for the activation reaction as indicated in the graph, and $K_{\rm d} = 1$ for the deactivation reaction. The maximum velocities of activation and deactivation reactions are given by $V_{\rm a} = V_{\rm d} = 100$, respectively.

where $E_{\rm a}$ is the signaling enzyme which switches inactive state Y to active state X, and $E_{\rm d}$ switches X off. Thus, the input signal is the concentration of $E_{\rm a}$. If each Michaelis-Menten reaction works near saturation, sharp response is obtained (Fig.8.6(b)).

In order to characterize the amplification of small changes in signal, we introduce the gain, which is defined as the ratio between the fractional change in the output signal X and the fractional change in the input signal S,

$$g = \frac{\Delta X/\bar{X}}{\Delta S/S}. \tag{18}$$

In this chapter, we consider that there is only a small change in S. Then, the gain is rewritten as $g = \frac{d\log X}{d\log S}$. For the Michaelis-Menten-type reaction and the push-pull reaction, the gain is plotted in Fig.8.6. Ultrasensitivity is defined as the response of a system that is more sensitive to change in S than is the normal hyperbolic response in Michaelis-Menten kinetics, in which the maximum gain g is unity. Thus, the maximum gain g of an ultrasensitive system is larger than unity.

In Fig.8.7, the gain is plotted as a function of the noise intensity. The gain and the noise intensity are numerically obtained independently. The gain is calculated by changing the intensity of the signal S, to $S + \Delta S$ and measuring the response ΔX in the output signal X from its stationary value $\bar{X}$. The noise intensity is calculated under the stationary condition without changing the signal intensity. Since the noise is not applied externally, but is intrinsically produced, this noise is called intrinsic noise. The numerical result clearly indicates that the gain is linearly proportional to the intrinsic noise intensity, i.e., $g \propto \sigma_{\text{in}}^2/\bar{X}$, where σ_{in}^2 is the variance of fluctuation in the concentration of X, and $\bar{X}$ is the average concentration.

This proportional relation is also obtained by analytical calculation and it is written as

$$g = \Theta \frac{\sigma_{\text{in}}^2}{\bar{X}} \tag{19}$$

where Θ is a particular constant depending on the reaction. Note that Θ is not a dimensionless parameter and depends on the measurement of X. For the derivation, see Ref. 20. Not only the push-pull reaction, but many types of reactions follows this gain-intrinsic noise relation Eq.19. For instance, the Michaelis-Menten type reactions, cooperative reactions such as allosteric enzyme models[28,29,33,34] and gene expressions. This relation tells us that a high response that is characterized by high gain results in large intrinsic noise, while large intrinsic noise implies high gain. The gain and the intrinsic noise cannot be controlled independently.

8.5. Propagation of Noise in Reaction Networks

No reaction works alone in cells. Many types of reactions interact with one another to form cascades and networks. If a signal, which is generated by some reaction, is noisy, the reaction regulated by the signal can be affected by the noise in the signal. Therefore, there are two noise sources; one is the noise inherent in its own reaction (intrinsic noise), and the other one is the noise generated due to the noise in the signal (extrinsic noise). Elowitz, et. al. first pointed out and demonstrated this distinction experimentally[4] in gene expression follow this relation. Theoretically, Paulsson analyzed the propagation of noise in gene network[14]. If a signal transduction reaction amplifies the small changes in the input signal, the noise in the input signal may also amplified. Here, we show how the amplification of noise is related to the gain[20].

When the signal S fluctuates with standard deviation σ_S, the fluctuation in the concentration X contains the extrinsic noise component. The standard deviation of the extrinsic noise is designated by σ_{ex}. The relative extrinsic noise intensity $\sigma_{ex}/\bar{X}$, is given by

$$\frac{\sigma_{\mathrm{ex}}}{\bar{X}} = g\sqrt{\frac{\tau_{\mathrm{S}}}{\tau + \tau_{\mathrm{S}}}}\frac{\sigma_{\mathrm{S}}}{S}. \tag{20}$$

where g is the gain, τ_S is the time constant of the noise in the input signal, and τ is the time constant of the signal transduction reaction. For the derivation, see Ref. 20. This gain-extrinsic noise relation indicates that the amplification rate of the input noise is at most the gain g. When the time constant of the input noise is large (slow noise), i.e., $\tau_{\mathrm{S}} \gg \tau$, the amplification rate of the input noise approaches the gain g. When the time constant of the reaction is much larger than the noise, i.e., $\tau_{\mathrm{S}} \ll \tau$, the noise in the input signal is averaged out and the amplification rate decreases proportionally to to $\sqrt{\tau_{\mathrm{S}}}$ as the time constant τ_{S} decreases. In this way, the amplification rate of the input noise depends on the time constants τ and τ_{S} as well as the gain g.

The total noise σ_{tot} is made up of the intrinsic noise σ_{in} and the extrinsic noise σ_{ex}, i.e., $\sigma_{\mathrm{tot}}^2 = \sigma_{\mathrm{in}}^2 + \sigma_{\mathrm{ex}}^2$. Therefore, from Eqs.(19) and (20), the total noise σ_{tot} is written as

$$\frac{\sigma_{\mathrm{tot}}^2}{\bar{X}^2} = g\frac{1}{\Theta\bar{X}} + g^2\frac{\tau_{\mathrm{S}}}{\tau + \tau_{\mathrm{S}}}\frac{\sigma_{\mathrm{S}}^2}{S^2} \tag{21}$$

where the first term on the right hand side is the intrinsic noise (Eq.(19)), and the second term is the extrinsic noise (Eq.(20)).

Since the intrinsic noise σ_{in} is proportional to the square root of g, and the extrinsic noise σ_{ex} linearly depends on the gain g, the dependence of the total noise σ_{tot} on the gain is both square root and linear of g. When g is small, the total noise dependence on g is square root of g, and when g is large, the total noise dependence is linearly proportional to g. Therefore, depending on the gain g, it is expected that there are two regions; the region in which the intrinsic noise is dominant in the total noise, and the region in which the extrinsic noise dominates the total noise.

For the push-pull reaction, the noises were calculated numerically by applying stochastic input signal. In Fig.8.8, the total noise intensity is plotted as a function of the gain. When g is small, the total noise σ_{tot}^2 linearly depends on the gain g. As the increase of g, the dependence of the total noise σ_{tot}^2 on the gain g approaches to square dependence on g. In this way, it is shown numerically that two kinds of regime exist; one is the intrinsic

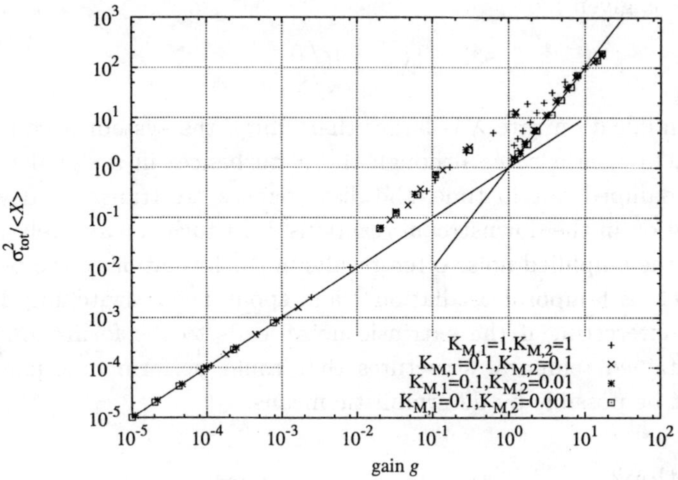

Fig. 8.8. Amplification of noise in signal transduction systems. In order to see the dependence of the total noise intensity σ_{tot} on the gain g, $\sigma_{\text{tot}}/\bar{X}$ is plotted as a function of the gain g. Changing the average concentration of the input signal, g, σ_{tot}, and $\langle X \rangle$ were obtained numerically. The numerical calculation was performed using the Gillespie's algorithm [23] as was the case in Fig.8.7. In the present case, the concentration of the input signal also fluctuates in time, and the average concentration increases under the condition that the relative noise intensity is maintained to be constant. The parameters: $V\text{a} = V_{\text{d}} = 10$, $K\text{a}$ and K_{d} are as indicated in the figure.

noise dominant regime and the other one is the extrinsic noise dominant regime.

Signal transduction systems that amplify signals also amplify the noise in the input signals. Consequently, the total noise is made up of the extrinsic noise as well as the intrinsic noise, as the relation Eq.(21) indicates. This relation is generalized in cascade reactions. In a cascade, a signal transduction system regulates another downstream signal transduction. Depending on the gain in each reaction, the output signal of such a cascade is dominated by the intrinsic or extrinsic noise. If the cascade consists of reactions with high gain, the extrinsic noise dominates the fluctuation in the output signal.

The amplification of the noise along a cascade implies the convective instability in dynamical systems with flux, typically found in fluid dynamics, traffic systems, and cascade reaction systems[35]. In the systems with convective instability, small disturbances are amplified as they are advected downstream. In the present case, from Eq.(20) , the amplification rate of noise $\lambda =$ is given by

$$\lambda = \frac{\sigma_{\rm ex}/\bar{X}}{\sigma_{\rm S}/S}. \tag{22}$$

If the amplification rate λ is larger than unity, the system is convectively unstable. In such a case, although the disturbances in a signal transduction are damped out in time, the disturbances are transmitted with the amplification in the downstream reactions[35]. In such convectively unstable systems, the amplified noise often results in the formation of temporal patterns, such as temporal oscillation[35] and spontaneous switching. It would be quite interesting if the extrinsic noise leads to the formation of such noise-sustained temporal structures that could perform functions, which could not be possible by deterministic means.

8.6. Outlook

The noise generate at some reaction propagates in the network of gene regulations and signal transduction. If a small change in a signal is amplified as it is propagated along the network, the signal processing ability of the system can be limited by large noise. In procaryote, the depth of a cascade of gene regulation is not so large. This is because these cascades are responsible for quick response to environmental variations. However, in eukaryote, the depth of a cascade of gene regulation and signal transduction is often large. From the viewpoint of noise, since the intrinsic noise generated in each reaction step is added to the noise propagated from the upstream reactions, long cascades may not be necessarily appropriate for amplification of small signal. One simple way to reduce the noise is to average the signal for sufficiently long time. However, in such a case, the quick response may be difficult, and sensitivity to a small change may be reduced.How signal transduction networks are designed to respond properly to noisy signals by using noisy molecular networks is an important future problem to elucidate the underlying 'design principle' of signal transduction network.

Acknowledgments

The author acknowledge stimulating discussions with Koichi Fujimoto at University of Tokyo and Alexander S. Mikhailov at Fritz Haber Institut.

References

1. H. Kuthan,*Prog. Biophys. Mol. Biol.* **75** 1-17 (2001).

2. A. Becskei, and L.Serrano, *Nature* **405**, 590-593 (2000).
3. E. M. Ozbudak, M. Thattai, I. Kurtser, A. D. Grossman, and A. Van Oudenaarden, *Nat. Genet.* **31**, 69-73 (2002).
4. M. B. Elowitz, A.J. Levine, E. D. Siggia, and P. S. Swain, *Science* **297**, 1183-6 (2002).
5. W. J. Blake, M. Kærn, C. R. Cantor, and J. J. Collins, *Nature* **422**, 633-7 (2003).
6. Y. Sako, S. Minoguchi, and T. Yanagida, *Nature Cell Biology* **2** (2000) 168-172
7. M. Ueda, Y. Sako, T. Tanaka, P. Devreotes, and T. Yanagida, *Science* 294 (2001) 864.
8. E. Korobkova, T. Emonet, J. M. Vilar, T. S. Shimizu, and P. Cluzel, *Nature* 428 (2004) 574-578.
9. T. Uyemura, H. Takagi, T. Yanagida, and Y. Sako, *Biophys. J.* (2005)
10. O. G. Berg, *J. Theor. Biol.* **71** (1978) 587.
11. H. H. McAdams, and A. Arkin, *Proc. Natl. Acad. Sci. USA* **94** (1997) 814-819
12. Thattai, M and Van Oudenaarden, A. (2001) *Proc. Natl. Acad. Sci. USA* 98 8614-8619
13. P. S. Swain, M. B. Elowitz, and E. D. Siggia, *Proc. Natl. Acad. Sci. USA* **99** (2002) 12795.
14. J. Paulsson, *Nature* **427**, (2004) 415-8.
15. F. Oosawa, *J. Theor. Biol.* **52**, (1975) 175-86.
16. F. Oosawa, *Bull. Math. Biol.*, **63**, (1975) 643-54,
17. P. B. Detwiler, S. Ramanathan, A. Sengupta, and B. I. Shraiman, *Biophys. J.* **79** (2000) 2801-2817.
18. O. G. Berg, J. Paulsson, and M. Ehrenberg, *Biophys. J.* **79** (2000) 1228-1236.
19. D. E. Koshland, Jr. *Science* **280**, (1998) 852-3.
20. T. Shibata and K. Fujimoto, *Proc. Natl. Acad. Sci. USA* **102**, (2005) 331-336
21. K. Sato, Y. Ito, T. Yomo, and K. Kaneko, *Proc. Natl. Acad. Sci. USA* **100** (2003) 14086.
22. N. G. van Kampen, *Stochastic processes in physics and chemistry* (North-Holland, Amsterdam ; New York) (1992).
23. D. T. Gillespie, *J. Phys. Chem.* **81**, 2340-2361 (1977).
24. D. T. Gillespie, (2000) J. Chem. Phys. 113 297
25. T. Shibata, Physical Review **E67**, (2003) 061906
26. T. Shibata, J. Chem. Phys., **119** (2003) 6629-6634
27. T. B. Kepler and T.C. Elston *Biophys. J.* **81** (2001) 3116
28. J. Monod, J. Wyman, and J. P. Changeux, *J. Mol. Biol.* **12**, 88-118 (1965).
29. D. E. Koshland, Jr., G. Nemethy, and D. Filmer, *Biochemistry* **5** (1966) 365-385.
30. E. R. Stadtman and P. B. Chock, *Proc. Natl. Acad. Sci. USA* **74** (1977) 2761.
31. A. Goldbeter and D. E. Koshland, Jr., *Proc. Natl. Acad. Sci. USA* **78** (1981)

6840.
32. C. Y. Huang and J. E. Ferrell Jr. *Proc. Natl. Acad. Sci. USA* **93**, 10078-83 (1996).
33. J. P. Changeux and S. J. Edelstein, *Neuron* **21**, (1998) 959-80.
34. U. Alon, L. Camarena, M. G. Surette, B. Aguera y Arcas, Y. Liu, S. Leibler, and J. B. Stock, *Embo J.* **17** (1998) 4238.
35. T. Shibata, *Phys. Rev.* **E 69**, (2004) 056218.

CHAPTER 9

AN INTELLIGENT SLIME MOLD: A SELF-ORGANIZING SYSTEM OF CELL SHAPE AND INFORMATION

Tetsuo Ueda
Research Institute for Electronic Science,
Hokkaido University, Sapporo 060, Japan
E-mail: ueda@es.hokudai.ac.jp

Dynamic aspects of biological information in the cell behavior and morphogenesis by the plasmodium of the true slime mold *Physarum polycephalum* are studied. Particular emphasis is put on the self-organization of cell shape and size in relation to intelligent functioning by the organism. The transduction of sensed information, integration of multiple signals, decision making, biocomputing, emergence of the rhythm etc. are shown related to a transition among dynamic dissipative structures on a metabolic network or among spatiotemporal order in a collective dynamics of coupled oscillators. Importance of global order in the amoeboid cell as a whole in cell functioning is clarified.

9.1. Introduction

Intelligence is the ability to perceive one's environment, to adapt to it and to work toward a goal. The behavior and morphogenesis are the appropriate reactions of the organism to its environment, and are ubiquitous among living organisms. Thus, every living organism from bacteria to higher animals has its own "intelligence". Among them, amoeboid cells are peculiar in that the organism changes its shape in a very complex way. This outward expression should be a reflection of inner state. By using characteristic feature of the plasmodium of the true slime mold, we analyze here the cell behavior and morphogenesis in the giant amoeboid cell of the true slime mold. The transduction of sensed information, integration of multiple signals, decision making, biocomputing, emergence of the rhythm, order among multiple rhythms, allometry in locomotion, etc. are described from a viewpoint of transitions among dynamic dissipative structures on a metabolic network or in a collective dynamics of coupled oscillators.[1,2,3,4,5]

9.1.1. *The True Slime Molds, Like Nothing on Earth*

The true slime molds are eukaryotic micro-organisms. They are unique not because they have solitary features in taxonomic classification, but because they share common features with all the other four kingdoms of fungus, protozoan, plant and animal. So the taxonomists more than a century ago named the organisms appropriately myxomycetes (slime + fungus), or mycetazoa (fungus + animal). Phytochromes, known as photosensors in higher plants, are also found in this organism, suggesting a link to plants. The organism reacts behaviorally, performing high degree of information processing like that of brain in higher animals. These latter two are our main topics in this chapter. The true slime molds are cosmopolitan and free-living fed on rotten logs and fallen leaves in forests all around the world. The rich variety in the morphology of sporangia and spores is a basis of classification of this group, and the number of species discovered is now reaching one thousand All share the common life cycle, as shown in Fig. 9.1. The myxoamoeba germinates from spores, and proliferates as a single nuclear cell by feeding with bacteria. Although all the myxoamoebae look similar, there is a sex differentiation. Two haploid amoebae with different sex types conjugate to form a diploid zygote. This is the beginning of the unusual life of plasmodium. Here unlike other living organisms, the nuclear division,

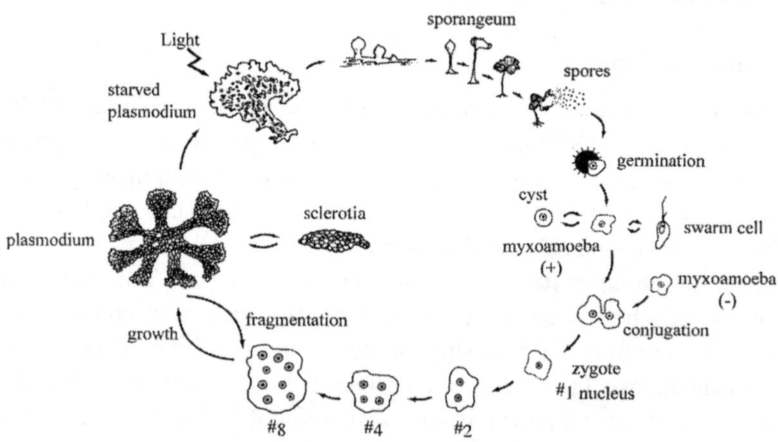

Fig. 9.1. The life cycle of the true slime mold *Physarum polycephalum.*

which takes place synchronously every ca..10 h, does not accompany the cell division. Thus the number of nuclei increases with 2^n after the nth nuclear division. In addition, when the two plasmodia with the same genetic fusion type encounter, they coalesce to form a single plasmodium. The lack of cell division and the natural cell fusion cause the plasmodium to become bigger and bigger, with an exception of the taxonomically primitive group of true slime molds, where the plasmodium divides itself while microscopic size. In our *Physarum*, it seems that there is no limit in size. So it is an everyday practice in our laboratory to have ten grams of wet plasmodium covering about 1 m^2 on agar gel, which contains about 10^{16} nuclei. Thus, the size of the plasmodium differs enormously, and yet the organism keeps seemingly a single cellular organization. Unlikeness of this organism emerges mainly from this fact, as we shall see later. When starved and irradiated with light, the plasmodium turns irreversibly into spores. When dried or subjected to high osmotic pressure, the plasmodium turns reversibly into a dormant form of sclerotia. The recently found morphogenesis is the plasmodial fragmentation where the plasmodium breaks into tiny spherical fragments with equal size, each containing about 8 nuclei. This phenomenon indicates that the plasmodium retains the ability of plasmodial (cell) division. This phenomenon is useful for studying the onset of the rhythm, memory effect and morphogen-like action of phytochrome, as discussed later.[6,7,8,9,10,11,12,13]

9.2. Cell Motility and Cell Behavior by the Plasmodium

9.2.1. *Cell Motility*

The plasmodium is a naked aggregate of protoplasm, and this living mass takes particular distribution patterns (cell shape) depending on external and internal conditions. When moving directionally, the fan-like sheet of protoplasm extends at the frontal region where the protoplasm becomes thick near the periphery, and intricate networks of veins develop toward the rear (Figs. 9.2a & 9.3). In the vein, protoplasmic sol moves to and fro very regularly and vigorously, the maximum flow velocity reaching more than 1 mm/sec. Note that this is the fastest protoplasmic streaming in the living world. The flow itself is passive, and is brought about by the difference in hydrostatic pressure. The ectoplasmic gel at any point in the organism exhibits active rhythmic contraction, and the coupling among the active elements should be responsible to bring about collective synchronization and hence coherent bulk flow. Like our muscle, the actin and myosin system is responsible for the contraction. However, unlike the muscle, the orga-

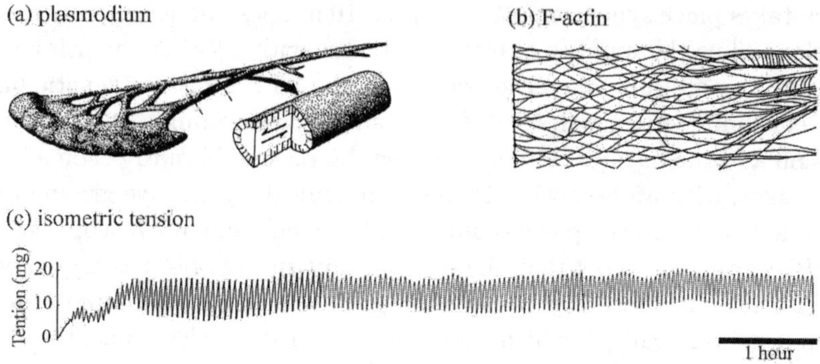

Fig. 9.2. The directionally moving plasmodium (a), the organization of actin filaments (b), oscillatory tension generation of the plasmodial strand (c).

nization of these proteins at the subcellular level is meticulously complex as seen in Fig. 9.2b. Out of this complex structure, global order emerges. When cut off from the mother plasmodium and hung on a tensiometer, the plasmodial strand exhibits the regular rhythmic contraction-relaxation cycle (Fig. 9.2c) and this oscillation is stable to external perturbation. Hence the contraction rhythm is considered as a limit cycle oscillation.[14,15,16,17]

9.2.2. *Chemotaxis*

The organism is not a simple motile machine, but is an information machinery. While migrating on the substratum where an attractant and a repellent are distributed alternately, the plasmodium avoids the repellent and moves toward the attractant regions, resulting in a zigzag trajectory (Fig. 9.3). This ability is called chemotaxis, and is known for more than a century. Soon after the discovery it was noticed that attractants were mostly, but not always, nutrients. This coincidence misled to a long-standing explanation of chemotaxis in terms of teleological arguments: the lower organisms moved toward attractants because these were beneficial as nutrients, and avoided repellents because they induced harmful effects. But around 1960s–70s, this theory was rejected by modern molecular biology: the microorganisms have sensors to receive environmental information with no regard to the effects which the stimulation induces afterwards. Take an example. We like sweet, not because it is nutrient, but just because it is sweet. We can only know by knowledge that same sweets are nutrients and cause dietary problems. All these are not known at the sensing level. Likewise, we dislike bitter, not

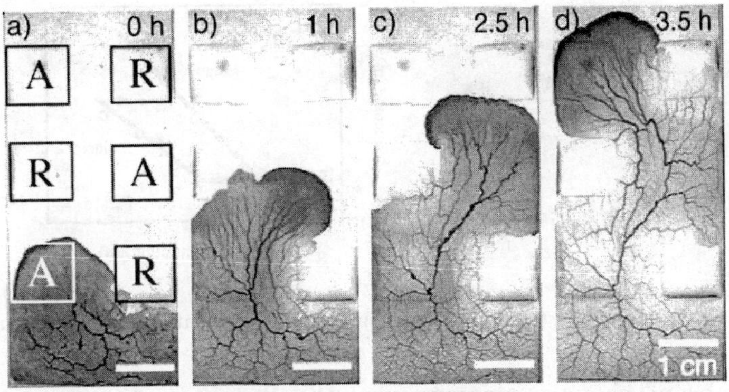

Fig. 9.3. The behavior of the plasmodium. The plasmodium migrates in a zigzag way by avoiding repellent regions. R: a repellent quinine, A: an attractant glucose.

because it might bring about harm for tissues, but just because it is bitter. The micro-organism has the similar sensing mechanism. The fact that chemicals which induce positive and negative chemotaxes coincides with those which induce beneficial and harmful effects respectively, has evolutionary origin: The organisms which happened to move toward nutrients had a better chance of survival than those moved away from nutrients, and in the long run the organisms with these characteristics have become the majority in the population.[18]

9.2.3. *Sensing and Transduction*

The study of chemotaxis by the plasmodium is, as discussed above, a search for making clear the following signal transduction pathways:

External stimulation → (a) Sensing → (b) integration → (c) judgment → (d) behavior

In higher animals, the sensing of chemical stimuli takes place at the surface receptor membrane of taste cell or olfactory cell. So is the case in micro-organisms. Note that the entire surface of the *Physarum* plasmodium is covered with the receptor membrane. The mechanism of chemoreception has been studied extensively by using microplasmodia; spherical plasmodia with about 100 μm diameter, which the *Physarum* plasmodium takes under submerged shake-culture conditions (Fig. 9.4a). For example, fluorescence probe methods reveal that the membrane increases its fluidity and becomes more hydrophobic upon recognition of chemical stimuli. Electrophoretic measurements show that the surface charge density of

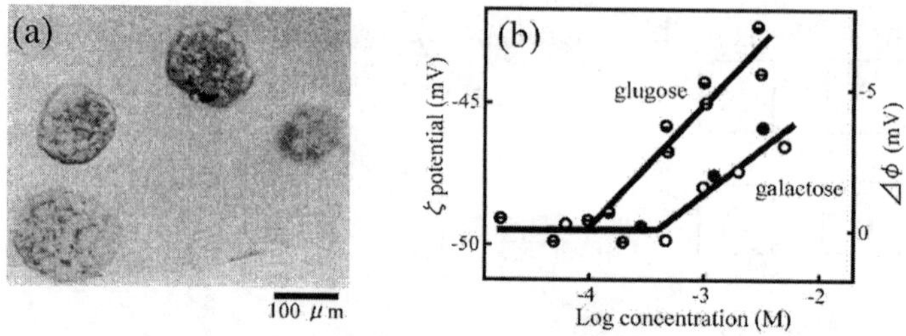

Fig. 9.4. Microplasmoda (a) and changes in surface potential in response to sugars (b).

the membrane changes accompanying the chemoreception, and that the changes in surface potential and transmembrane potential agree with each other for any chemicals such as salts, sugars, and hydrophobic substances (Fig. 9.4b). The receptor potential increases monotonously with increase in external concentration above respective thresholds. Thus, the changes in the membrane conformation upon chemoreception are involved in the sensory transduction from sensing to intracellular signals.[19,20,21,22,23,24,25,26,27,28]

9.2.4. *Search for Second Messengers*

Candidates for the intracellular signaling substances, i.e. second messengers, are those which vary the concentration upon chemoreception. On this criterion many intracellular chemical components are searched for second messengers, the experiment which the large size of the plasmodium is beneficial. The concentrations of cAMP, cGMP, ATP, pH and NADH are found to vary upon chemical stimulation, and increase and decrease of these chemical components correspond to positive and negative chemotaxis, respectively. Although details are not yet known, we note here for the following discussion that the quality and quantity of external stimuli are transduced to intracellular signals at the receptor membrane.[29,30,31,32,33,34]

9.3. Integration of Sensed Information in Chemotaxis

The organisms in nature are exposed to multiple stimuli at a time, and yet perform an appropriate behavior. This indicates that the organism is able not only to sense the single stimulus, but also to integrate the multiple sensed signals. How does the organism respond to a mixture of conflicting stimuli of an attractant and a repellent? Surprisingly, the results are not additive: the presence of an attractant causes the avoidance reaction to the repellent to occur at higher concentrations. Quantitatively, 10-fold increase in the attractant concentration needs 10-fold higher concentration of the repellent for negative chemotaxis. On the other hand, the membrane potential responds at the similar concentrations as those stimulated with the repellent alone. Thus, the changes in the recognition threshold at the behavioral level are not ascribed to the receptor membrane, but to the downstream in the sensory transduction pathway; i.e., "integration". In the following we analyze further the stimulus-response relationships, and propose a chemical model for the integration in terms of competitive binding. Experimental results so far obtained are summarized as follows: To a single stimulus, the behavioral response R is expressed as a function of the concentration C of the external chemical stimulus.

$$R = R_0 C/(1 + KC), \tag{1}$$

where, K is the constant expressing the sensitivity and differs for different stimuli. The response R may be the percentage of the responded organisms among the whole population, and takes positive or negative values for attractants and repellents, respectively (Fig. 9.5). Among the attractants are some monocarbohydrates and some aminoacids, and among the repellents are the bitter substances and the acids.[35,36]

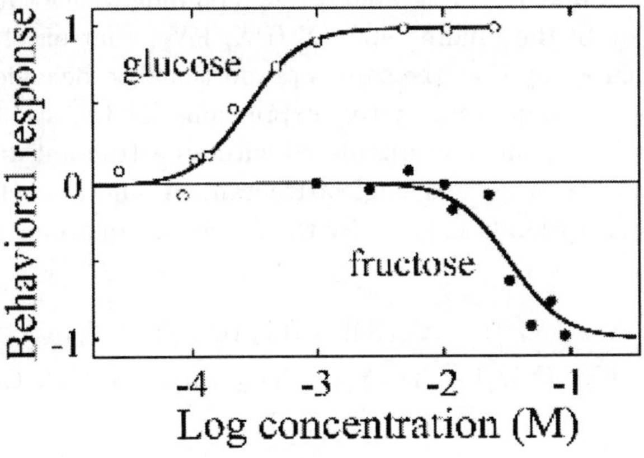

Fig. 9.5. Behavioral response of the *Physarum* plasmodium to various chemicals. Data can be best fit with $R = R_0 C^2/(1 + (KC)^2)$ rather than Eq. 1.

9.3.1. *A Model for Integration*

The stimulus-response relationship to a chemical as expressed by Eq. 1 varies in the presence of the other chemical, and we will consider the mechanism which accounts for the experimental results. Our model consists of three steps; (a) signal transduction, (b) integration and (c) link to cell behavior, and assumes the followings.

(a) Signal transduction to intracellular chemicals: Attractants and repellents induce two different kinds of intracellular signaling substances, X and Y, respectively. The intracellular concentrations are considered, for the sake of simplicity here, proportional to external concentration C above the chemoreception thresholds.

(b) Signal integration: Both X and Y bind competitively with the same substance P. The reaction scheme is:

$$\mathrm{X} + \mathrm{P} \rightarrow \mathrm{PX} : \mathrm{Y} + \mathrm{P} \rightarrow \mathrm{PY}$$

(c) Link to the control of behavior: The summation of the fraction of P bound to X and Y is proportional to the cell behavior. i.e., R = PX-PY (minus sign is due to negative taxis). This model is generalized to any number of multiple stimulations: The summation of the external stimuli $\Sigma_i \mathrm{C}_i$ is transduced to internal signal $\Sigma_i(\mathrm{X}_i+\mathrm{Y}_i)$ at the receptor membrane. Intracellular signal X and Y is integrated through binding with a substance P and the summation of the binding state $\Sigma_i(\mathrm{PX}_i\text{-}\mathrm{PY}_i)$ controls the motile machinery in such a way that the sum is proportional to behavioral response. Our model gives some quantitative expressions for the stimulus-response curves when the organism is stimulated with an attractant and a repellent at a time. The responses to a single attractant, A, and a repellent, B, at an external concentration C is given by the following equation.

$$R = K_{\mathrm{X}} \cdot (\mathrm{X})/[1 + K_{\mathrm{X}}(\mathrm{X})] = K_{\mathrm{XA}}(\mathrm{C_A})/[1 + K_{\mathrm{XA}}(\mathrm{C_A})], \tag{2}$$

$$R = K_{\mathrm{Y}} \cdot (\mathrm{Y})/[1 + K_{\mathrm{Y}}(\mathrm{Y})] = K_{\mathrm{XB}}(\mathrm{C_B})/[1 + K_{\mathrm{YB}}(\mathrm{C_B})], \tag{3}$$

where the parentheses indicates the concentration, and K_{X} and K_{Y} are equilibrium constants defined as

$$K_{\mathrm{X}} = (\mathrm{PX})/[(\mathrm{X}) \cdot (P)], \; K_{\mathrm{Y}} = (\mathrm{PY})/[(\mathrm{Y}) \cdot (P)]. \tag{4}$$

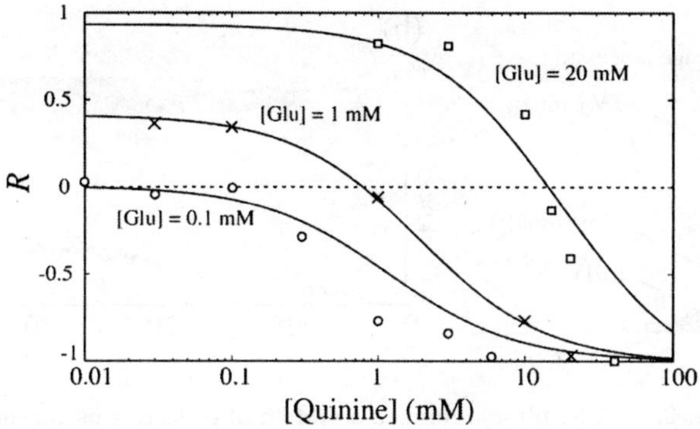

Fig. 9.6. Chemotactic response to quinine in the presence of glucose.

The response to a mixture of an attractant and a repellent is given as follows:

$$\begin{aligned} R &= P_2/P_0 - P_1/P_0 \\ &= [K_\mathrm{X} \cdot (\mathrm{X}) - K_\mathrm{Y} \cdot (\mathrm{Y})]/[1 + K_\mathrm{X}(\mathrm{X}) + K_\mathrm{Y}(\mathrm{Y})] \\ &= [K_\mathrm{XA}(\mathrm{C_A}) - K_\mathrm{XB}(\mathrm{C_B})]/[1 + K_\mathrm{XA}(\mathrm{C_A}) + K_\mathrm{YB}(\mathrm{C_B})], \qquad (5) \end{aligned}$$

where P_0 is the total concentration of the substance P. Fig. 9.6 shows the theoretical predictions by eq. 5 together with experimental results. Both theoretical and experimental results agree with each other, indicating that the present model is likely. Particularly the response becomes the function of the ratio of the concentrations A and B at high concentrations.

9.4. Collective Dynamics of Coupled Oscillators in Cell Behavior

Let us recall that the *Physarum* plasmodium exhibits the regular to-and-fro streaming of protoplasm (sol, endoplasm) which is brought about by the active rhythmic contraction of the protoplasm (gel, ectoplasm) taking place everywhere in the plasmodium. This indicates that the plasmodium is a system of coupled oscillators, and that the cell behavior as well as the protoplasmic streaming should be regarded as a coherent collective dynamics of this system. This expectation is realized experimentally with use of a sheet-like 2D extension of the plasmodium.

The organism is illuminated from below with weak IR light, and is observed from above with a IR sensitive video camera. The wavelength for

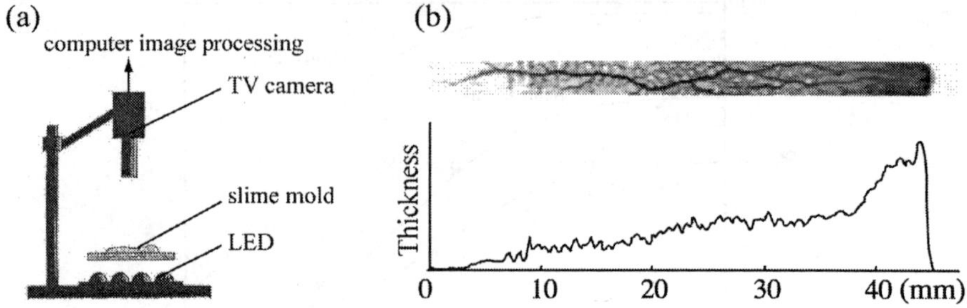

Fig. 9.7. The plasmodium as a system of collective oscillators.

observation light should have as little effects as possible on the plasmodium, and the action spectra for photoavoidance and photomorphogenesis show that IR region (~900 nm) has little effects on the plasmodium (see Fig. 9.24). Assuming that the plasmodium absorbs or scatters the light uniformly, we may derive the Lambert-Beer like relation that the logarithm of the relative brightness level I^* decreases linearly with the thickness l of the plasmodium: $\log I^*/I_0^* \propto l$. This relationship is confirmed experimentally. So, we can measure the spatial distribution of the thickness all over the plasmodium. In a directionally migrating plasmodium, the plasmodium becomes most thick near the advancing front (see Fig. 9.7b). The thickness oscillates everywhere in the plasmodium accompanying the shuttle streaming as well as the active contraction. To pick up the oscillating components near the intrinsic rhythm of the shuttle streaming, the difference in the brightness level is taken. $I(x,t) = I^*(x,t) - I^*(x,t-\tau_0)$, where I^* is the brightness level, and τ_0 is taken about one third of the intrinsic oscillation. The intrinsic rhythm is not limited to the one related to the shuttle streaming. In later section we show that the plasmodium exhibits multiple rhythm, and discuss how these rhythms form a global order.[37,38,39,40,41,42,43]

9.4.1. *The Response to External Stimulation*

Upon local stimulation of the plasmodium, the phases of the oscillation at the stimulated part become antiphase with those at the unstimulated regions (Fig. 9.8). These phase shifts are induced similarly upon stimulation of both attractant and repellent above the threshold concentration. This fact indicates that phase shift does not represent the discrimination of

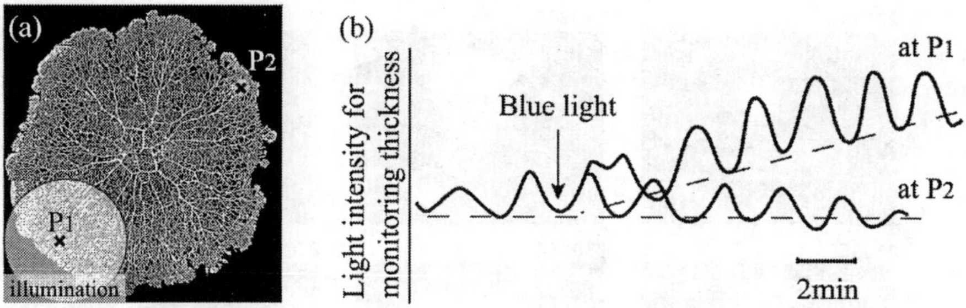

Fig. 9.8. Phase shift induced by local stimulation.

stimulants (attractants and repellents), but simply express the recognition of something stimulated.

In a concentrically expanding plasmodium, oscillation patterns are radially symmetric, and the waves propagate from the periphery to the inside, in most cases. Upon local stimulation, a new wave begins to propagate around the stimulated region. The response to local high temperature stimulation is shown in Fig. 9.9. The wave propagates from the stimulated region to the surroundings. The propagation of phase wave is expressed as $I(x,t) = I(kx + \omega t)$, where k is the phase gradient vector, ω is the frequency of the oscillation. The phase gradient vector k is calculated to be $k = \omega\, \mathrm{grad} I/(\mathrm{d}I/\mathrm{d}t)$, and is determined experimentally for various stimulation. When stimulated with attractants such as high temperature, oatflakes, glucose etc., the k vectors orient to point away from the stimulated region. On the other hand, the k vectors orient to point toward the region when stimulated with repellents such as low temperature, UV and blue light, and high salt concentrations. The fact that the phase gradient vectors classify attraction and repulsion uniquely indicates that the integration of the sensed information has a dynamic spatio-temporal characteristics observed at the level of interacting oscillators. When the entire region of the organism is stimulated uniformly, the frequency of the oscillation is slightly modified. The oscillation becomes faster and slower for attractants and repellents, respectively. It seems common in a system of nonlinear oscillators that the faster oscillator entrains the slower one, and that the phase gradient vectors point from the entraining to the entrained regions. The phase wave and phase gradient vectors observed in the slime mold suggest that a similar mechanism is working in the plasmodia.

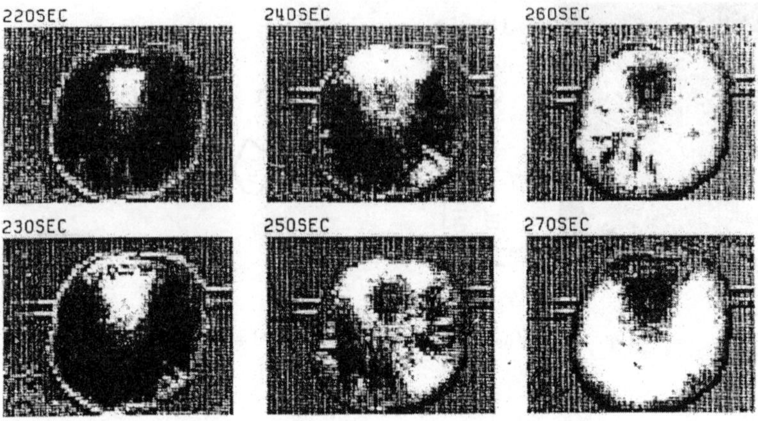

Fig. 9.9. Propagation of phase wave upon local stimulation.

9.4.2. *Alteration of the Judgment by Oscillatory Stimulation through Entrainment*

A new question arises whether the orientation of phase gradient vector really controls the behavior. To show this, new evidence is needed. Nonlinear oscillators in general entrain to external oscillations in a certain range of frequency and amplitude, and so is the *Physarum* oscillator. The intrinsic rhythm of the plasmodium entrains to the external temperature oscillation, and the phase gradient vectors orient as expected from the general nonlinear oscillators. The plasmodium is subjected locally to a low temperature which oscillates a little bit faster than the intrinsic rhythm. The phase gradient vectors orient to point away from the stimulated region, and the thickness there becomes thicker and thicker. Thus, the low temperature stimulation turns into an attractive stimulus even if the temperature is kept low on the overage. Similarly the high temperature turns into repellent stimulation when oscillated a little bit slower than the intrinsic rhythm, where the phase wave vector orients to point toward the stimulated region. Thus, we have the following transduction pathway:

Stimulation→orientation of phase gradient vector→behavior
↑ (entrainment)
external oscillation

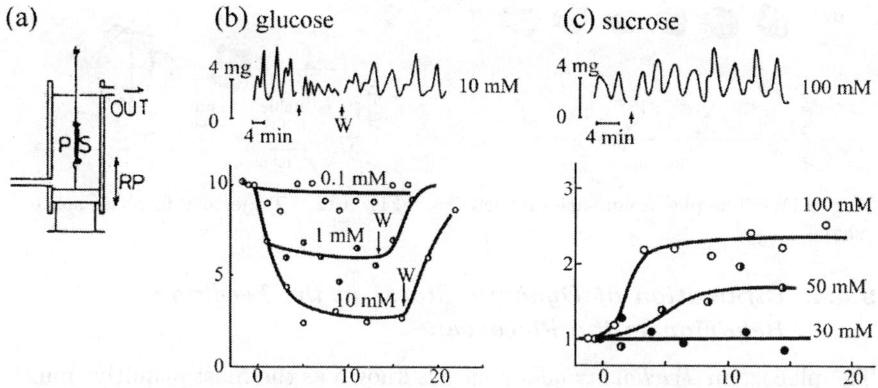

Fig. 9.10. The isometric tension in response to chemical stimulation.

9.4.3. *Correspondence of Tactic Behavior with Contractility*

Positive and negative taxes by the *Physarum* plasmodium are also correlated to the viscoelastic properties of the plasmodium. A plasmodial vein when hung on a tensiometer exhibits a contraction and relaxation cycle on the tension. Effects of chemicals on the isometric tension are summarized as follows. All the attractants and repellents tested induce the relaxation and contraction, respectively. Thus, the cell behavior is reduced to such mechanical properties as contraction and relaxation. The stronger contraction on the average brings about the difference in hydrostatic pressure to squeeze out the protoplasmic sol from the stimulated region, and hence the negative taxis results. Similarly in the reverse-wise, the relaxation at the stimulated region induces the flow of protoplasm into the stimulated region, and hence positive taxis results.

We have at least two correlates for positive and negative taxes; one is the propagation of phase wave, and the other contractility. But these mechanisms cannot explain the pseudopodium formation, i.e., the extension of the periphery. When the air is replaced with nitrogen, the locomotion stops but the rhythmic streaming continues. This fact indicates that the extension and contraction are separate functions in the plasmodium. The coupling and coordination mechanism among them is yet to be studied.[44,45,46,47,48,49,50]

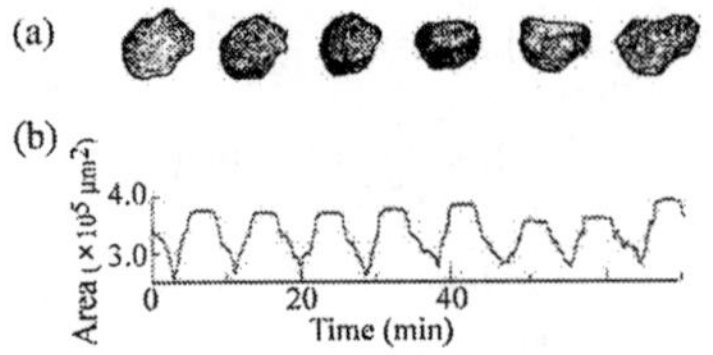

Fig. 9.11. The placozoan varies its cell shape.

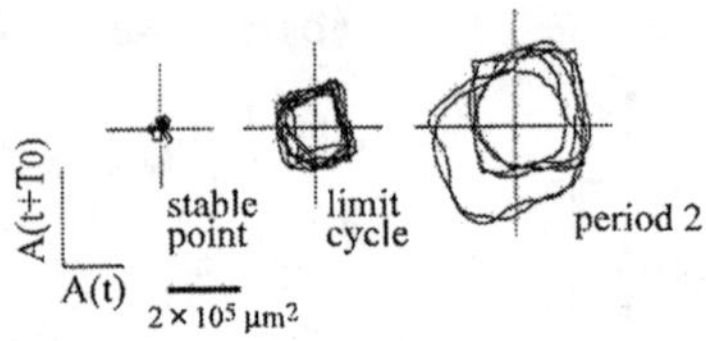

Fig. 9.12. Trajectory in phase space.

9.4.4. *Bifurcation of Dynamic States in the Feeding Behavior by the Placozoan*

The placozoan *Trichoplax adhaerence* is known as the most primitive multicellular organism. The organism swims on the surface of substratum by beating cilia which cover the whole organism. Yet the flat plate-like organism changes its shape accompanying the locomotion. The addition of food material (yeast extract) induces a periodic variation in cell shape, and the limit cycle oscillation turns into the period 2 in the higher concentrations. Thus, bifurcations from stable state to limit cycle, and then limit cycle to period 2 occur in the feeding behavior in the placozoan. Taking into account the general feature of bifurcation of this kind in nonlinear dynamics, it is rather strange that the stationary state in the plasmodium is not noticed.[51]

9.5. Chemical Oscillations as a Basis for the Rhythmic Contraction

The contractile apparatus made of actin/myosin system changes its structure periodically accompanying the rhythmic contraction. The G-F transformation of actin seems involved where the actin-modulating proteins participate. Biochemically the activities of these modulators are shown to be controlled by free calcium and ATP. Physiologically it has also shown that the concentrations of these modulators oscillate accompanying the rhythmic contraction.

Metabolic oscillation, however, is not limited to the modulators of the actin/myosin system. Regulators of metabolic pathways such as NADH, cAMP, cGMP, pH, also oscillate accompanying the rhythmic contraction. As an example, the variations of NADH and cAMP concentrations are shown in Fig. 9.13. The frequency of the metabolite oscillation is the same with the rhythmic contraction, but the phases differ for different metabolites. The ATP and calcium concentrations oscillate in-phase and antiphase, respectively, with the contraction-relaxation cycle, while the membrane potential and cGMP concentration oscillate with 90 degree different phases. In this way, the protoplasm of the plasmodium is in an oscillatory state as a whole. That is, the notion of hoemeostasis is not applicable in this organism.[52,53,54,55]

9.6. Transition of Chemical Patterns Accompanying the Selection of Cell Behavior

The cell behavior and hence cell shape can be classified into several patterns, suggesting that the underlying spatio-temporal dynamics exhibits stationary states. So we may expect typical spatial patterns for respective cell behavior. The large size of the plasmodium allows us to measure the spatial distribution of intracellular chemical components such as ATP, cAMP, cGMP, NADH etc. Here is an experimental protocol: (I) Freeze the plasmodium rapidly with liquid nitrogen. (II) Cut off the whole plasmodium lattice-wise into 10 x 10 pieces. (III) Extract the intracellular chemicals. (IV) Analyze the amount of chemical components as well as the total protein.

Fig. 9.14 summarizes typical patterns in the distribution of ATP concentration when the organism takes various cell shape. When the organism is migrating directionally, the ATP concentration is high at the frontal part and decreases toward the rear (a). When the organism is extending concentrically, the ATP distribution looks like a round trough, being high at the peripheral region (b). When the organism is extending like a fan, the distribution looks like the oyster shell, still on the average being high at the advancing front (c). When the organism is placed in a trough where only one narrow channel is open to outside, the organism extends and covers the entire confinement, and then rushes to go through the channel, the

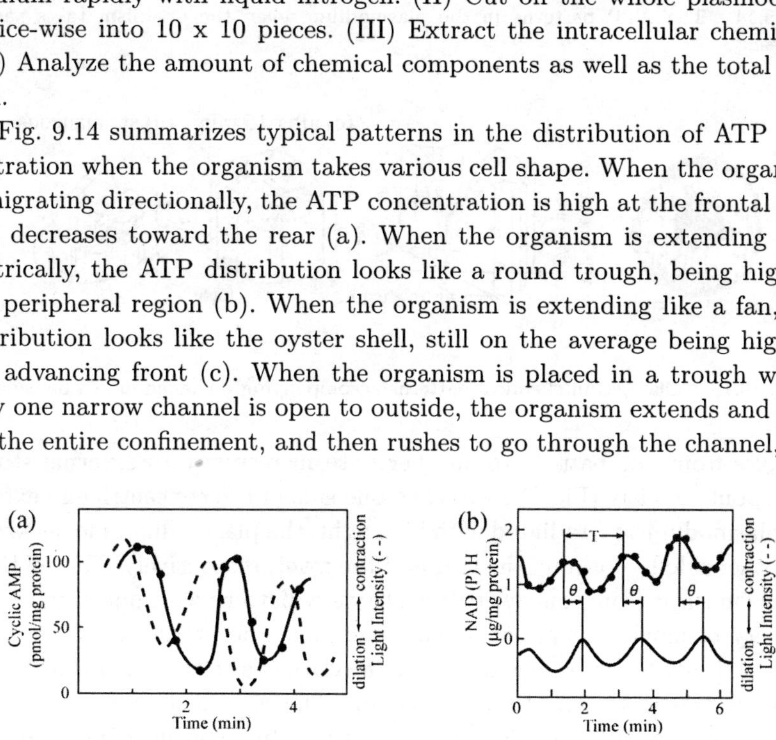

Fig. 9.13. Chemical oscillations accompanying the rhythmic contraction.

frontal part becoming thick. The ATP distribution becomes strongly polar, the frontal part being high and rear part being undulated (d). When the organism is confined in a square, the small veins develop all through the occupied area rather uniformly. Yet the ATP distribution exhibits three peaks and three valleys with 3-rotational symmetry (e). The ATP distribution

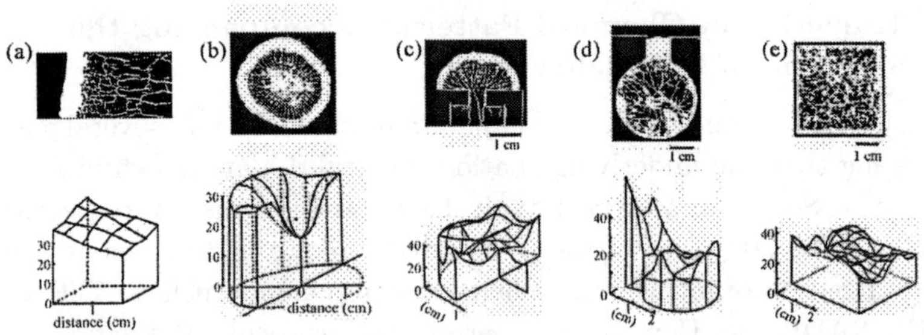

Fig. 9.14. The ATP patterns in the plasmodium when the organism takes various shapes.

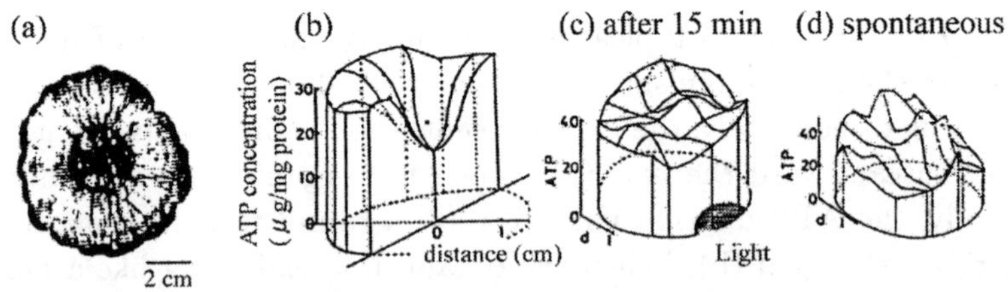

Fig. 9.15. Changes in chemical patterns accompanying a change in cell behavior.

changes from one pattern to another both in response to external stimuli and spontaneously (Fig. 9.15). When one side of the concentrically extending plasmodium is irradiated with blue light, the plasmodium moves toward the other side by forming the rear and the front, respectively. The ATP distribution turns from concave pattern to wavy pattern with polarity (c). The polarity of movement appears spontaneously. The organism only extends concentrically to some size, and after the critical size the organism exhibits polarity, mostly moving to one direction. The ATP distribution is a wavy pattern with polarity, similar to that induced by external stimulation (d). Note that the cell shape still look similar excepting that the frontal region becomes thinner.[56,57,58,59,60,61,62]

9.6.1. *Theory of Cell Behavior in Terms of Dissipative Structure*

Oscillations and pattern formation of intracellular chemical components as described above are explained in terms of a reaction-diffusion coupling model. For example, the well-known theoretical model of the Brusselator, and both experimental and theoretical study of glycolytic system are within this framework. The system is open to environment, and consists of a network of chemical reactions which are coupled with diffusion. As a computer simulation and experimental study demonstrate, this chemical system can take on various dynamic states depending on both kinetic parameters and boundary conditions. Prigogine has named these spatio-temporal order as dissipative structures, to emphasize the open nonequilibrium situation far away from equilibrium as distinct from those formed in equilibrium situation. This notion should be applied to the cellular system in a somewhat extended form. Namely, the cell behavior is governed by the chemical patterns, and the selection of the behavior is nothing but the transition from one pattern to another.

9.6.2. *Link to the Organization of Cytoskeleton and Chemical Pattern*

The cell shape is supported by the organization of cytoskeleton. So the above theory for cell behavior is only finalized after showing the correspondence between chemical patterns and the organization of cytoskeleton. The cytoskeleton in the *Physarum* plasmodium is seen as birefringent fibrils under a polarizing microscope, and the distributions of the fibrils and ATP concentration are compared (Fig. 9.16). At first the organism is extended uniformly and the polarity is then induced by applying a dc electric field. Soon after the stimulation, the ATP concentration decreases just inside the anodal side and becomes higher toward the cathodal side, with undulations in between (c). The birefringent fibrils become abundant just inside the anodal side and decrease toward the cathode, with undulations in between (d). So, roughly speaking, birefringent fibrils become abundant where ATP concentration decreases. In other words, self-organization of chemical patterns may cause reorganization of the cytoskeleton, and hence cell behavior.

This mechanism of cell behavior gives a clear answer why the cell is able to behave coordinately and also as an integrated unity, a question that the previous theories of amoeboid cell motility could not solve. Based on the theory of self-organization of chemical patterns, the chemical pattern itself

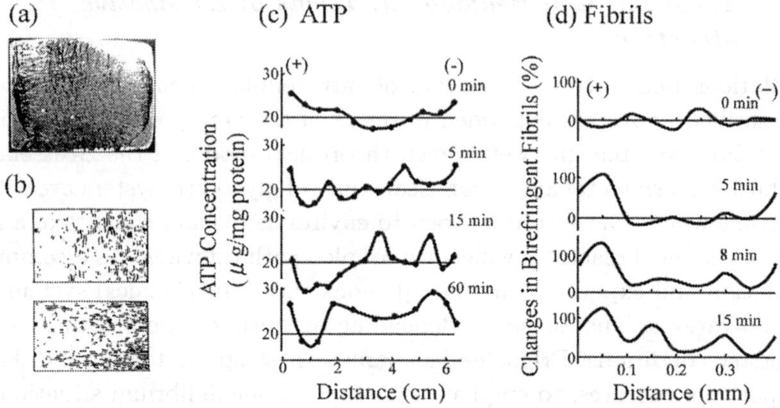

Fig. 9.16. Birefringent fibrils under a polarizing microscope (a) and the distribution of the fibrils under a dc field on comparison with chemical pattern (b).

is generated by both kinetic parameters and boundary conditions of the system. In other words, the spatial pattern itself is nothing but a reflection of the system as a whole. For full understanding of the cell behavior, we have to know what selection mechanism acts on this basic scheme. The integration mechanism as discussed in the section 9.3 should be extended to include a link with the selection mechanism.

9.7. Computing by Changing Cell Shape

As seen above and as its Greek name implies (*amoibe* = change), the plasmodium changes its shape apparently freely in response to changing environments. Instead of asking the molecular mechanism, we now study the informational aspects of these shape changes. Let us consider the follow-

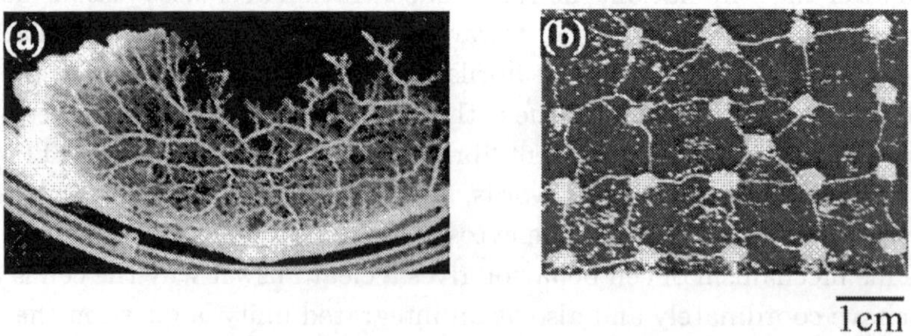

Fig. 9.17. Cell shape at free migration (a) and on distributed food (b).

ing two cases: (1) the plasmodium is allowed to migrate on plain agar, and (2) the plasmodium is allowed to move around freely on a Petridish where flakes of oatmeal are placed lattice-wise (Fig. 9.17). In the former case, the plasmodium migrates by extending like a sheet at the frontal part and by developing intricate networks of veins toward the rear (a). In the latter case, the protoplasm covers the food material and the protoplasm elsewhere disappears except thick veins, mostly each one connecting the shortest food site (b). In the same situation, we may pose global, informational and mathematical questions: (1) Determine a distribution of protoplasm which allows the known amount of protoplasm seek food and move around most effectively. (2) Determine the distribution of protoplasm where the known amount of protoplasm makes as much contact with food material as possible to take in nutrients and at the same time the intracellular communication becomes more effectively. We may expect that the cell shape in each case described above is an expressed solution to the two mathematical problems.

In this way, having no definite form does not mean that the organism has no sufficient information to build its own form, but does mean that the organism has the ability to take into environmental information to determine its own form. In other words, the organism reads the situation, and makes a "computation", and expresses the solution as its own form. In the following we show explicitly that the plasmodium does compute and solve common mathematical puzzles.[63,64,65,66]

9.7.1. *Solving a Maze Problem*

The organism is confined in a maze. The amount of protoplasm is so adjusted as to cover the entire region almost uniformly by developing thin veins (Fig. 9.18a). Then the sufficient food is placed at two appropriate

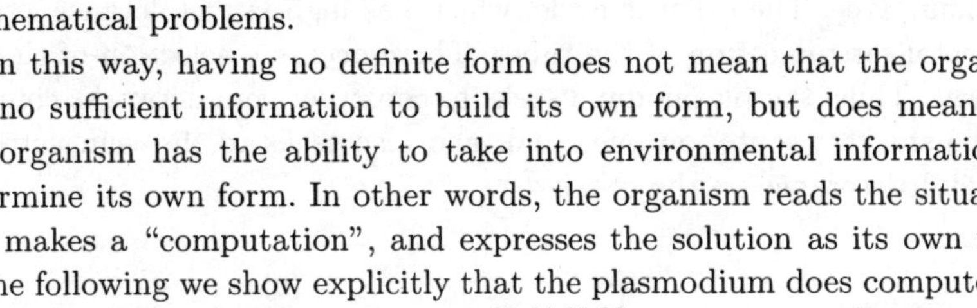

Fig. 9.18. Maze-solving by the *Physarum* plasmodium.

places in the maze. The plasmodium moves to cover the food, and the distribution of protoplasm in the maze begins to change. At first, the protoplasm at the dead ends disappears, and thick veins are formed at all the possible routes connecting the two food sites (b). Then, among the two connections the shorter route remains while the longer one disappears (c). This is the way how the plasmodium solves the maze problem.

9.7.2. *Solving the Steiner Problem*

The appropriate amount of the plasmodium is confined on the circular disk to distribute uniformly. Food material is added at three sides near the periphery to make the vertices of a triangle. The mass of protoplasm flows to the food sides, leaving veins connecting them. Two common patterns appear. One is the tubular network connected them via a short pathway, which is analogous to the mathematically shortest route known as Steiner's Minimum Tree. The other is cycle, which has high fault tolerance against accidental disconnection of the tubes. There occurs a selection of the two patterns. Thus, the plasmodium seeks to construct vein networks connecting the shortest route, on one hand, and also seeks to the vein networks with fault tolerance on the other.

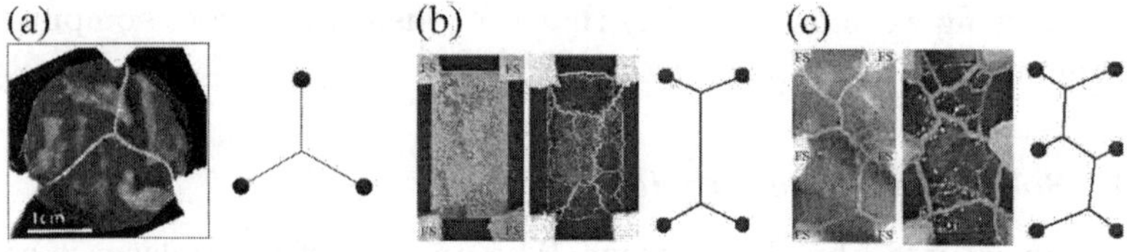

Fig. 9.19. Shape of the plasmodium for multiple food sources.

9.7.3. *Formation of Veins by External Oscillation*

The vein of the plasmodium is a structure which allows the protoplasmic sol to flow effectively. Since the flow is brought about by a difference in hydrostatic pressure, the volume flow obeys the Poiseuille equation: The total volume V of fluid flowing through the cylindrical tube per second for an incompressible fluid is given as

$$\mathrm{d}V/\mathrm{d}t = \pi(P_1 - P_2)R^4/8L\eta, \tag{6}$$

where R is the radius of the cylinder, L the length, η the viscosity of the fluid, $P_1 - P_2$ the difference in the hydrostatic pressure through the tube. The pressure difference is generated by the difference in local oscillation, and hence becomes large when the oscillation becomes antiphase.

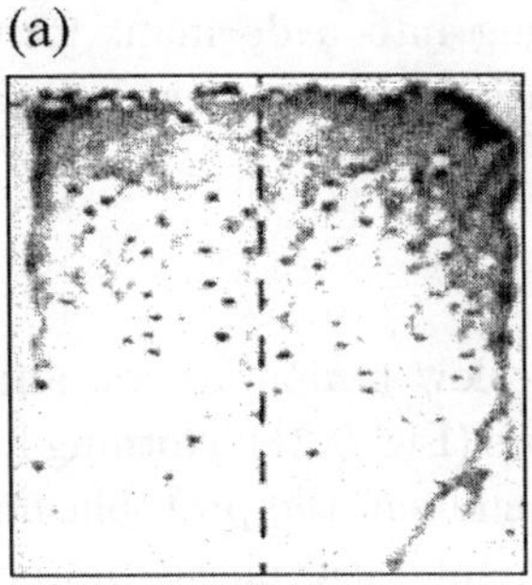

Fig. 9.20. Induction of vein formation at the frontal region by external oscillation (a); control, (b); under lateral anti-phase oscillation.

In directionally moving plasmodium, phase difference appears along the direction of movement and the thick veins are formed in this direction, peripheral and inner part oscillating in antiphase. When temperature is varied laterally on one half and the other oscillating in antiphase, the intrinsic oscillation entrains to the external one and becomes antiphase laterally. Then a while later, thick veins are formed laterally near the frontal part. Thus, the veins are flexible structure which changes its position as well as diameter according to the oscillation patterns. This result has a general important meaning. We take it that the vein is a structure, and that the flow is a function. Commonly the function is explained in terms of structure: e.g., The flow becomes faster as the vein becomes thicker. The above result indicates the reverse case that function affects structure. Since the oscillation patterns contain the global information, this dynamic coupling between function and structure is a basis for intelligent formation of veins as shown above.

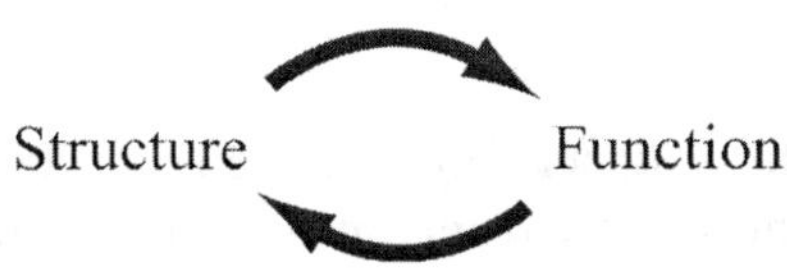

9.8. Fragmentation of the Plasmodium: Control of Cell Size

The *Physarum* plasmodium becomes larger and larger without limit when external conditions are favorable for growth. Recently it is found that there is a natural mechanism to limit the size of the plasmodium; that is, the plasmodium breaks temporarily into pieces either by cooling or by irradiation with light without transforming into a dormant form. This new phenomenon of plasmodial fragmentation gives a new look and tool for studying the plasmodial organization.[67,68,69,70]

9.8.1. *Thermo-Fragmentation*

Thin-spread plasmodium is exposed to low temperatures, and changes in cell shape are monitored for several hours (Fig. 9.21). Nothing happens until 4 h after the stimulation. And all of a sudden, the protoplasmic streaming stops and the plasmodium breaks into spherical pieces in about an hour. Due to this long latency this phenomenon might be unnoticed till recently. Some characteristics of thermo-fragmentation are summarized below. Irrespective of the temperature, the fragmentation takes place at around 5 h after cooling below a certain critical temperature (about 15°C). Thus, the time of the fragmentation is independent of temperature. In 10 to 15 h after stimulation, the plasmaodial fragments coalesce to resume a single large plasmodium. The addition of cyclohexhimide (an inhibitor of protein synthesis) and actinomycin D (an inhibitor of transcription) suppresses the fragmentation, indicating that the new expression of genes as well as protein synthesis are necessary for the fragmentation.

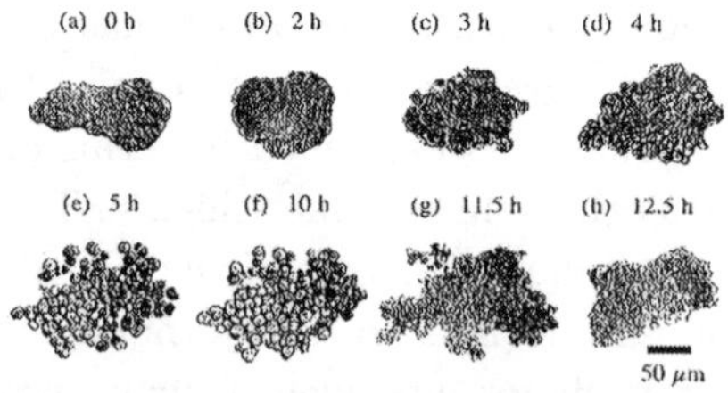

Fig. 9.21. A time course of the plasmodial fragmentation.

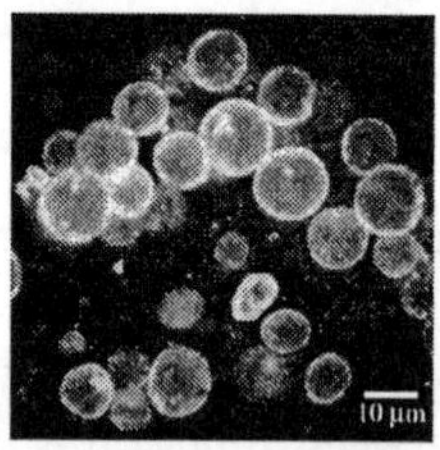

Fig. 9.22. Actin organization in plasmodial fragments.

To summarize, the fragmentation consists of the following pathways: **Sensing → (a) signal transduction → (b) gene expression→ (c) protein synthesis → (d) morphogenesis.**

Thus, the fragmentation is a new morphogenesis, being different from spherulation and sporulation. Each plasmodial fragment contains 8 ± 2 nuclei. The size becomes halved under compression or under aerobic conditions. Thus, the size distribution exhibits a bistability. Interestingly, these numbers 8 and 4 fall on a series 1, 2, 4, 8..., which appear by synchronous increase in the number of nuclei after zygote formation. Hence, some interaction among nuclei is suspected. The plasmodial fragmentation might have evolutionary origin. In the primitive true slime mold, such as *Echinostelium minutum*, the plasmodium does not grow large, but divides itself when nuclei reach a certain number. Thus, the fragmentation might be an ancestral expression of plasmodial division. Genetic analysis of the fragmentation will clarify genes which control the size of the plasmodium.

9.8.2. *Photofragmentation and its Photosystem*

The fragmentation of the plasmodium is also induced by irradiation with light. The time courses and morphological characteristics are similar to that induced by temperature. The active wavelengths are UVB, blue and far red (750 nm), as seen in Fig. 9.24. The induction period, after which the stimulation is no longer needed, is about half an hour. In the long wavelengths, antagonistic effects are found: The far red-induced fragmentation is inhibited by red light (670 nm) photoreversibly. Thus, one of the photosensors for photofragmentation is a phytochrome. The fluence rate-response for the induction with far red light as well as inhibition by red light is quantitatively explained by the following model: (a) A photoreceptor takes two forms of P_r and P_{fr}; (b) P_r converts to P_{fr} and *vice versa* upon irradiation with red and far-red light, respectively; (c) P_r converts to P_{fr} under dark; and (d) the fraction of P_r is proportional to the response, i.e., photofragmentation. Thus the kinetics for a change in P_{fr} is as follows:

$$\mathrm{d}[\mathrm{P_f}]/\mathrm{d}t = K_{\mathrm{fr}} I_{\mathrm{fr}} [P_{\mathrm{fr}}] - (K_{\mathrm{r}} I_{\mathrm{r}} + K_{\mathrm{d}})[P_{\mathrm{r}}], \tag{7}$$

where

$$[\mathrm{P_r}] + [\mathrm{P_{fr}}] = [\mathrm{P_0}]. \tag{8}$$

At photoequilibrium (Eq. 7) is solved to give

$$[\mathrm{P_r}]/[\mathrm{P_0}] = R = 1/\{1 + (K_{\mathrm{r}} I_{\mathrm{r}} + K_{\mathrm{d}})/K_{\mathrm{fr}} I_{\mathrm{fr}}\}. \tag{9}$$

This model predicts that linear relationships hold among $1/R$ vs $1/I_{\mathrm{fr}}$ under constant I_{r}, and $1/R$ vs I_{r} under constant I_{fr} for induction and inhibition, respectively. This is confirmed experimentally. Thus the photoconversion

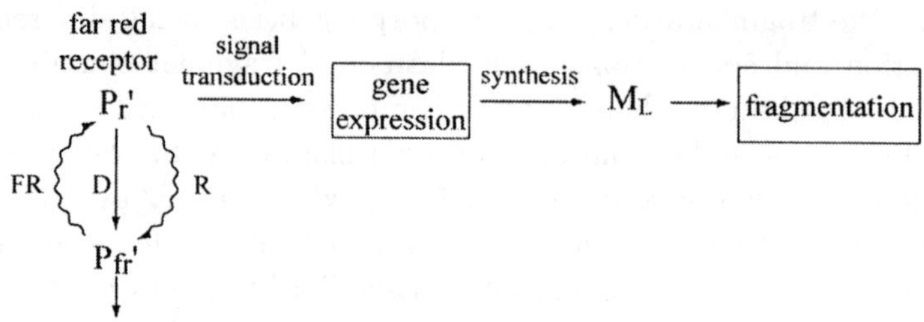

Fig. 9.23. Photosystem for plasmodial fragmentation.

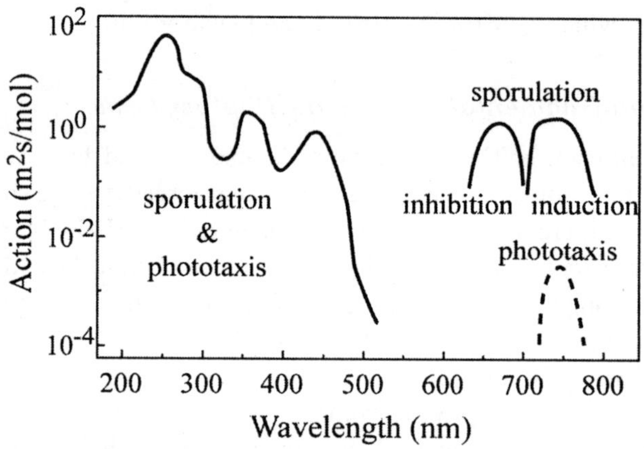

Fig. 9.24. Action spectra for sporulation and phototaxis.

model at the photoreceptor is applicable for the fragmentation of the plasmodium. Contrary to the phytochrome in higher plants, the red form is active in signal transduction.

The presence of photoconvertible pigments is demonstrated photobiochemically. Difference absorption spectra indicate the absorption change similar to the action spectra at red and far-red regions, as seen in the action spectra shown in Fig. 9.24.

9.9. Memory Effects and Morphogen: Phytochrome as Morphogen in the Fragmentation

A new phenomenon comes into light serendipitously in unsuccessful trials for determining the induction period in the thermo-fragmentation. There is no induction period in thermo-fragmentation; that is, the interruption of low temperature treatment at any time before 4 h fails to induce the fragmentation. However, when the low temperature treatment is resumed, the fragmentation takes place much earlier than 5 h. In other words, there is a memory effect. Moreover, this memory effect is affected by the irradiation with red and far red light in an antagonistic way.

Here is an experimental protocol to characterize this memory effect quantitatively. The plasmodium is exposed to low temperature (T_1) at time zero, and transferred to high temperature (T_2) at time t_1, and resumed to the low temperature at time t_2. Observe and measure the time when the plasmodium fragments into pieces. Red and far red light is irradiated where necessary.

Experimental parameters are : the duration of the first stimulation $\tau_1 = t_1$, the duration of the interruption of the stimulation $\tau = t_2 - t_1$, and the duration of the second stimulation before the fragmentation $\tau 2$.

The results are summarized as follows:

(a) The memory M is defined as $M = 5 - \tau_2$, and M decays exponentially with the interruption time τ. $M = M_0 \exp(-k\tau)$.
(b) The red light irradiation (pulse) at any time resets the memory of the low temperature treatment.
(c) The far red light recovers the memory which is decayed in the dark and reset with red light.
(d) Effects with red light and far red light are photoreversible.
(e) Inhibitors of transcription and translation suppress the fragmentation.

Our experimental results indicate that there exists a substance M which determines the morphogenesis when the concentration increases above a critical value, and that this morphogen is the Pr form of phytochrome. The reaction scheme is shown below:

The experimental result (a) indicates that the morphogen M is formed at a constant rate A under stimulation, and that the M decays with the first order kinetics. Thus,

$$dM/dt = A - kM. \tag{10}$$

The experimental results (b)-(d) indicate that the M is a P_r form of phytochrome which changes into inactive P_{fr} form thermally and with red light irradiation. The experimental result (d) indicates that there is a photocycle

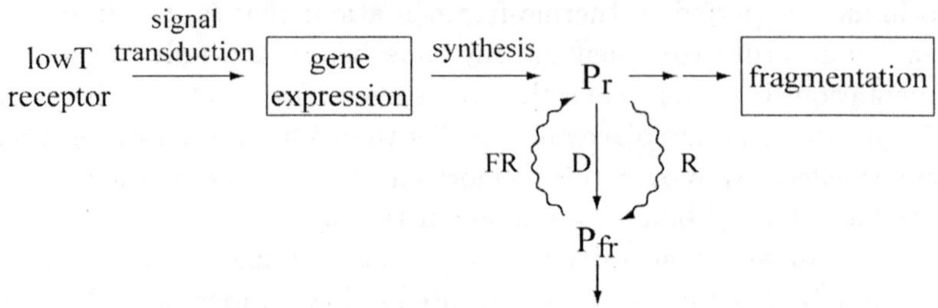

Fig. 9.25. Phytochrome acting as morphogen.

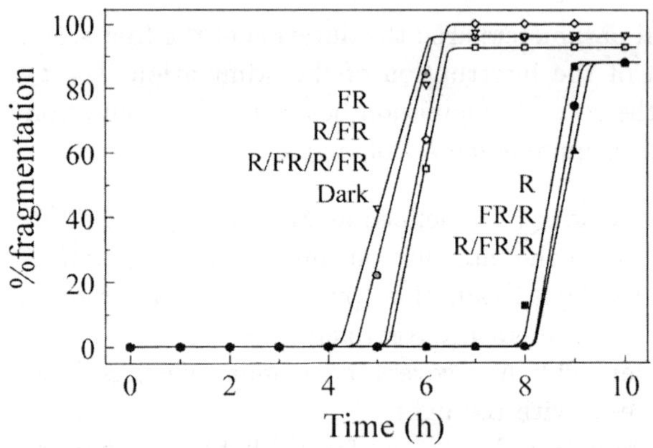

Fig. 9.26. Photoreversibility of the red and far-red light in the fragmentation.

between P_r and P_{fr}.

$$\mathrm{d}M/\mathrm{d}t = A - kM - k_{\mathrm{R}}MI_{\mathrm{R}} + k_{\mathrm{FR}}M^*I_{\mathrm{FR}}, \tag{11}$$

$$\mathrm{d}M^*/\mathrm{d}t = kM + k_{\mathrm{R}}MI_{\mathrm{R}} - k_{\mathrm{FR}}M^*I_{\mathrm{FR}} - k_{\mathrm{d}}M^*. \tag{12}$$

The above chemical kinetics explain quantitatively all the results obtained experimentally. Particularly the effects of red and far red light are photoreversible (Fig. 9.26). The pulses of red and far-red light irradiated successively induces the fragmentation where the light irradiated at the last time determines the effect. Note that the phytochrome acts as a morphogen as well as a photoreceptor (Fig. 9.25).[71]

9.10. Locomotion

9.10.1. *Allometry in Locomotion Velocity*

The large plasmodium migrates much faster than the smaller one. When the plasmodium is allowed to move in a narrow trough, the locomotion and hence the distribution of the protoplasm becomes one dimensional (See also Fig. 9.7b). The locomotion velocity V increases linearly with the total mass of the plasmodium M and decreases with the width of the trough L with the slope 0.36 in double logarithmic scale as shown in Fig. 9.27: Thus, the following empirical relationship holds:

$$V \propto (M/L)^{1/3} \tag{13}$$

The mechanism for this allometric relationship is yet to be studied.[72]

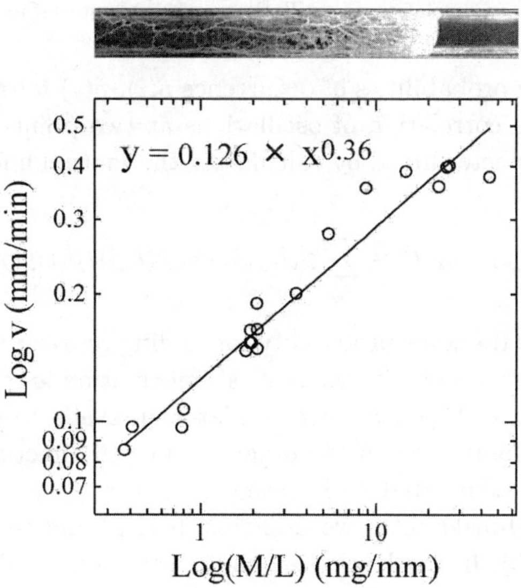

Fig. 9.27. The dependence of the locomotion velocity on the total mass of the plasmodium.

9.10.2. *Correlation of Oscillations During Directional Movement*

The locomotion in the above condition seems stationary, and hence the time courses of the oscillation over 2 hours at a point are digitized, say

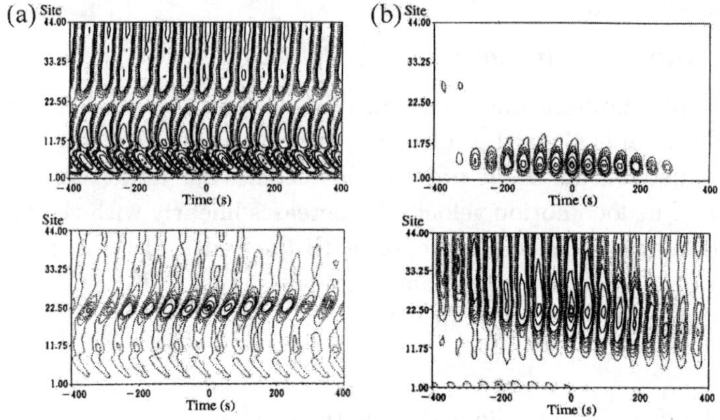

Fig. 9.28. Correlation of oscillations analyzed mutual information.

three levels. The probabilities of occurrence $p(i)$, $p(j)$ for marks i and j are determined. The correlation of oscillations at two points x_1 and x_2, at a time interval t is determined by calculating the mutual information defined as follows:

$$I(x_1 : x_2, t) = \sum_{i,j} p(i, j, t) \log[p(i, j)/p(i)p(j)], \tag{14}$$

where $p(i, j, t)$ is the joint probability for finding marks i and j at the time interval t. A contour map for the mutual information is shown in Fig. 9.28.

The strong correlation among oscillators prevails from the tip to the tail of the plasmodium when the organism moves directionally (a). When the organism is stimulated with oatflakes at the tip portion, this wholesome coherence breaks into two domains, frontal and rear parts correlating separately (b). In the later case, the plasmodium stops moving. Thus, the some correlation over the whole organism is necessary for the strong locomotion.[73]

9.11. The Emergence of the Rhythmic Streaming

So far, we take it implicitly that the plasmodium at any condition exhibits vigorous and regular shuttle streaming. However, the taxonomists noticed long before that this property does not appear for a while after the zygote formation where the plasmodium remains tiny (see Fig. 9.1), and got the view that the shuttle streaming is related to the maturation, as often

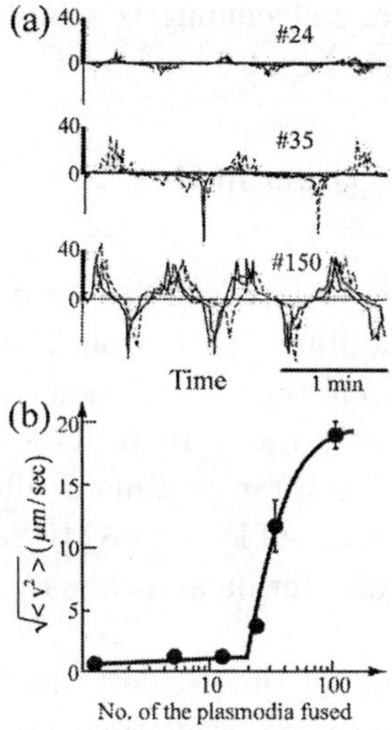

Fig. 9.29. Size dependence of the locomotion velocity.

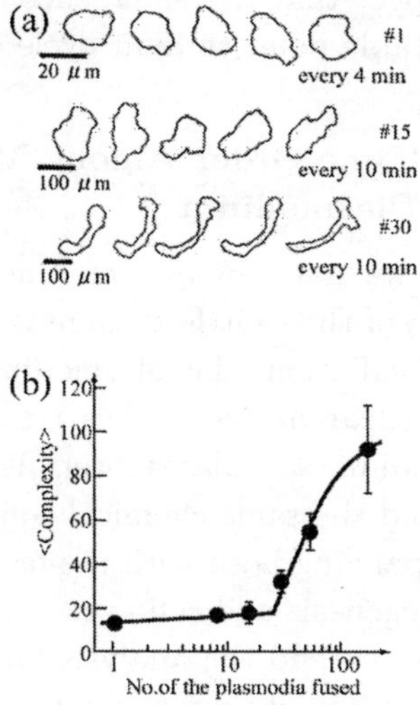

Fig. 9.30. Cell shape as a function of cell size.

seen in many living organisms. One of the other possible explanations is physicochemical; A certain amount of protoplasm is necessary for the shuttle streaming. Normally the growth and the maturation of the plasmodium take place inseparably. However, the new phenomenon of the plasmodial fragmentation gives new experimental opportunity; We can prepare the plasmodia with various size having the same age. Here is an experimental procedure: (a) Prepare the thin-spread plasmodium, and induce the fragmentation by cooling. (b) Leave the given number of plasmodial fragments. (c) Raise the temperature, and wait until single tiny plasmodium appears after coalescence. (d) Observe the changes in cell shape and the protoplasmic streaming in the plasmodia. The results with the plasmodia having various size but the same age are shown in Figs. 9.29 & 9.30. The protoplasmic streaming becomes rhythmic and vigorous only when the plasmodium becomes larger than the critical size of about 20 fragments = 160 nuclei. The onset of the rhythmic streaming corresponds to changes in cell shape. The changes in cell shape are quantified by a dimensionless parameter complexity defined as Complexity = (periphery)^2/(area). The cell shape remains rather round when the plasmodium is tiny. The cell becomes elongated rather strongly when the size exceeds a certain value. The critical values for cell shape and the onset of the streaming agree with each other.

We expect that the emergence of the rhythmic streaming is a bifurcation from stable point to limit cycle oscillation (see Fig. 9.12).[74]

9.12. Time Order Among Multiple Rhythms in the Plasmodium

So far we have focused on the narrow time scale which covers only the rhythm of the shuttle streaming in the plasmodium. Apart from this short-term oscillation, the plasmodium exhibits long-term oscillations as well; The nuclear division takes place synchronously every 10 h. The starved plasmodium sporulates intermittently every 5 h after continuous illumination, and the same chemical components such as ATP and cAMP oscillate accompanying both with the fast rhythmic contraction and slow rhythm of morphogenesis and cell cycle.

Here we extend the analysis for a wider range of time scale to include all the rhythmic phenomena with short- and long-term oscillations, and show that some global order emerges in the plasmodium.[75,76,77,78,79,80]

9.12.1. *Long-Term Changes in Cell Shape of the Plasmodium*

Changes in cell shape are monitored for a long time under very dim illumination. The organism looks like a silhouette. The contour line of the cell is traced every 30 sec for over 20 hours, and the peripheral length is calculated. The time series data are analyzed with a microcomputer. Typical example of changes in peripheral length is shown in Fig. 9.31. On the slow and large variations are superimposed the slower and larger variations, on which much slower and much larger variations are superimposed.

The multiple rhythms coexist in the changing patterns of cell shape, and the plasmodium exhibits 7 oscillations altogether. The results are summarized in Table 1, together with other periodic phenomena.

9.12.2. *Multiple Oscillations*

Oscillations with characteristic time scales are separated by averaging procedures, and the oscillatory components together with some characteristics are shown in Fig. 9.32. With period T_6 (= 4 h), the organism takes rather round and elongated shape alternately. With the period T_5 (=30 min), the periphery becomes uneven and smooth alternately. With the period T_3 (= 1.5 min), the plasmodium blows and shrinks slightly and rather uniformly.

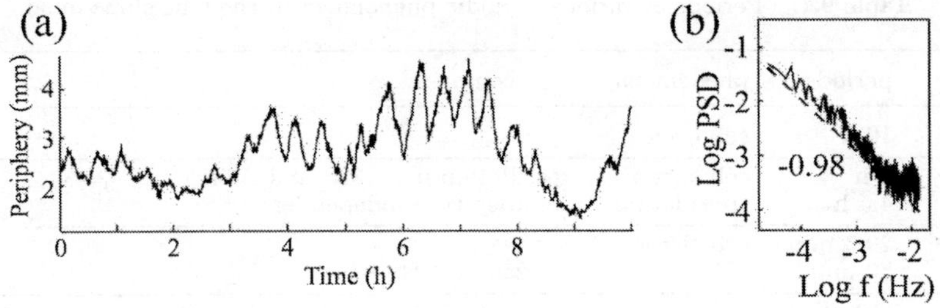

Fig. 9.31. The long-term variation in changes in peripheral length during free movement of the plasmodium (a) and the power spectral density (b).

Table 9.1 summarizes all the oscillatory phenomena found in this organism. The periodic phenomena are numbered in order of shorter periods. There are 7 periodicities altogether, and phenomenon No.4 is skipped for the sake of simplicity. By taking the ratio of periods between nearby and the next nearby oscillations, we find that the periods vary in a manner of geometric progression:

$\mathbf{T_{i+1}/T_i = 7}$ **and** $\mathbf{T_{i+2}/T_{i+1} = 3}$ **where i = 1, 3, 5.**

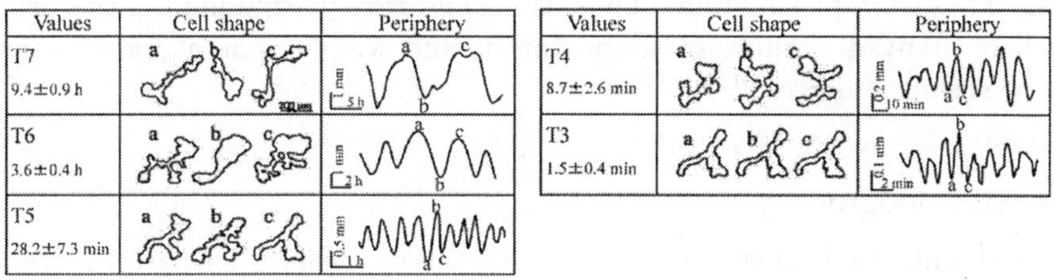

Fig. 9.32. Oscillatory components and characteristics in changes in cell shape.

9.13. Concluding Remarks as Future Prospects

We have shown experimentally that the *Physarum* plasmodium exhibits intelligence comparable with that of our own, by extensive use of the peculiar characteristics of the organism. Our approach will now combine with the other fields of disciplines, particularly with genomics and modeling. The National Human Genome Research Institute, one of the National Institute of Health, USA, added 18 organisms to sequencing pipeline in August 2004, and among them is our *Physarum*. We may expect very soon that

Table 9.1. Periods of various periodic phenomena in the true slime mold.

	periods	phenomena	comments
T_7	10 h	cell cycle	
T_6	4 h 4.5 h	cell shape sporulation	oscillation in cAMP and ATP, temperature independent
T_5	29.3 min 30 min	cell shape	 in 250 mM EtOH
T_4	(10 min)		oscillations in phase relationship
T_3	1.3 min 1.6 min	cell shape contraction	component independent of size, oscillations in cATP, ATP, Ca^2+, NADH and H^+ temperature independent
T_2	22.5 sec	cell shape	tiny plasmodium
T_1	3.3 sec	cell shape	

T_2/T_1=6.8 T_4/T_3=6.8 $T_6/T_5 = 6.7$
T_3/T_1=24.2 T_5/T_3=23.2 $T_7/T_5 = 20.3$

$T_{i+1}/T_i = 7$ and $T_{i+2}/T_{i+1} = 3$; $i = 1, 3, 5$

the whole genome of this organism is at our hands. The other kind of approach is modeling. The pioneer is the biophysics group in Russia, and this kind of approach should be extended to include the intelligent behavior of the *Physarum* plasmodium. Thus, the *Physarum* plasmodium is the place where many disciplines of science meet, and will become a cradle where new science emerges.[81,82]

Acknowledgments

I would like to dedicate this paper to my three teachers: The late Professor Noburo Kamiya (1913-1999), without being fascinated by his lecture on cell motility to the 3rd year students of the Science Department at the Osaka University and without his and his groups' lasting encouragement, I could not even attend the *Physarum* study.: The late Professor Wohlfarth-Bottermann (1923-1997), without his kind acceptance of my stay at the University of Bonn as a research fellow of the Alexander von Humboldt Foundation at the earliest year of my academic carrier, I could not foster my interest as an international researcher in the interdisciplinary field: The late Professor Yonosuke Kobatake (1926-1988.10.31), without being introduced to the field of non-equilibrium thermodynamics and without togetherness with him for 20 years as a student and coworker, I could not enjoy and continue my research. I thank Professor A. S. Mikhailov and Professor S. Müller for inviting me to the Workshop at Berlin and also to the Ottovon-Guericke-Universität at Magdeburg on December 2003; my second visit to Germany after a quarter of a century. Thanks are due to every member of our research group, Dr. Toshiyuki Nakagaki, Dr. Seiji Takagi and Mr.Koji Suzuki for stimulating discussions and for preparing the manuscript.

References

1. A. Perz-Miravete. *Behaviour of Micro-organisms*, (Plenum Pub.Comp.Ltd. 1973).
2. K.W. Jeon. *The Biology of Amoeba*, (Academic Press, New York, 1973).
3. J. Loeb. *Forced Movements, Tropisms, and Animal Conduct*, (Dover, 1973).
4. M. J. Carlile, ed. *Primitive Sensory and Communication Systems*, (Academic Press, London, 1975).
5. E. Stahl. *Bot. Zeit.* **10**, 145 (1884).
6. A. Lister and G. Lister. *A Monograph of the Mycetozoa*, (The British Museum, 1925).
7. W. D. Gray and C. J. Alexopoulos. *Biology of the Myxomycetes*, (The Ronald Press Company, New York, 1968).
8. C. J. Alexopoulos and C.W. Mims. *Introductory Mycology*, (John Wiley & Sons, New York, 1979).
9. Aldrich, H.C., and Daniel, J.W. *Cell Biology of Physarum and Didymium*, (Academic Press, New York, 1982).
10. M. J. Carlile and S. C. Watkinson. *The Fungi*, (Academic Press, London, 1994).
11. K. Baumann. *Naturfilme*, (Lindenstrasse 40, 72810 Gomaringen, Germany, 2003).
12. A. Huttermann. *Ber. Deutsch. Bot. Ges.*, **86**, S.1 (1973).
13. *PHYSARUM NEWSLETTER* 35:1 (2003).
14. N. Kamiya. *Protoplasmatologia* **8a**, 1. (1959).
15. W. E. Wohlfarth-Bottermann. *J. Exp. Biol.* **81**, 15. (1979).
16. W. Majani, W. Stockem, K. E. Wohlfarth-Bottermann, M. Osborn and K. Weber. *Eur.J.Cell Biol.* **28**, 103 (1982).
17. R. D. Allen and N. Kamiya, ed. *Primitive Motile Systems in Cell Biology*, (Academic Press, New York, 1964).
18. T. Ueda and Y. Kobatake. In *Cell Biology of Physarum and Didymium*, Ed. H.C. Aldrich and J.W.Daniel. (1982), p. 112.
19. M. Hato, T. Ueda, K. Kurihara and Y. Kobatake. *Biochim. Biophys. Acta* **426**, 73. (1976).
20. T. Ueda and Y. Kobatake. *Cytobiologie* **16**, 16. (1977).
21. K. Terayama, T. Ueda, K. Kurihara and Y. Kobatake. *J. Membrane Biol.* **34**, 369. (1977).
22. T. Ueda and Y. Kobatake. *J. Membrane Biol.* **34**, 351. (1977).
23. T. Ueda and Y. Kobatake. *Cell Structure and Function* **3**, 129. (1978).
24. M. Ataka, A. Tsuchii, T. Ueda, K. Kurihara and Y. Kobatake. *Biochem. Physiol.* **61A**, 109. (1978).

25. T. Ueda and Y. Kobatake. *Biochim. Biophys. Acta* **557**, 199. (1979).
26. T. Ueda, T. Hirose and Y. Kobatake. *Biophys. Chem.* **11**, 461. (1980).
27. T. Ueda and M. J. Carlile. *J. Gen. Microbiol.* **129**, 2475. (1983).
28. T. Ueda, M. Hasumoto, T. Akitaya and Y. Kobatake. *Protoplasma* **124**, 219. (1985).
29. T. Hirose, T. Ueda and Y. Kobatake. *J. Gen. Microbiol.* **121**, 175. (1980).
30. T. Hirose, T. Ueda and Y. Kobatake. *Cell Struc. Function* **5**, 281. (1980).
31. H. Satoh, T. Ueda and Y. Kobatake. *Cell Struc. Function* **7**, 275. (1982).
32. T. Akitaya, T. Hirose, T. Ueda and Y. Kobatake. *Gen. Microbiol.* **130**, 549. (1984).
33. T. Ueda, Y. Mori, T. Nakagaki and Y. Kobatake. *Photochem. Photobiol.* **47**, 271. (1988).
34. T. Ueda, Y. Mori, T. Nakagaki and Y. Kobatake. *Photochem. Photobiol.* **48**, 705.(1988).
35. S. Takagi and T. Ueda. *at the Ann. Meet.Biophys. Soc.*, (Japan, 2004).
36. D. J. Knowles and M. J. Carlile. *J. Gen. Microbiol.* **108**, 17.(1978).
37. K. Matsumoto, T. Ueda and Y. Kobatake. *J. Theor. Biol.* **122**, 339. (1986).
38. K. Matsumoto, T. Ueda and Y. Kobatake. *J. Theor. Biol.* **131**, 175. (1988).
39. T. Nakagaki and T. Ueda. *J. Theor. Biol.* **179**, 261. (1996).
40. T. Nakagaki, H. Yamada and T. Ueda. *Biophys. Chem.* **82**, 23. (1999).
41. Y. Kobatake, T. Ueda and K. Matsumoto. In *Biosensors and Biocomputers*, ed. F. Hong, (Plenum Press, 1990), p. 223.
42. T. Ueda, K. Matsumoto and Y. Kobatake. In *Molecular and Biological Physics*, ed. by K.S. Mishra, (Kluwer Academic Publishers, Dordrecht, 1990), p. 133.
43. T. Ueda. In *Oscillations and Morphogenesis*, ed. L. Rensing, (Marcel Dekker, 1993), p. 147.
44. T. Ueda, K. Terayama, K. Kurihara and Y. Kobatake. *J. Gen. Physiol.* **65**, 223. (1975).
45. T. Ueda, M. Muratsugu, K. Kurihara and Y. Kobatake. *Exp. Cell Res.* **100**, 337. (1976).
46. M. Hato, T. Ueda, K. Kurihara and Y. Kobatake. *Cell Struc. Func.* **1**, 155. (1976).
47. T. Ueda and K. Goetz von Olenhusen. *Exp. Cell Res.* **116**, 55. (1978).
48. T. Ueda, K. Goetz von Olenhusen and K. E. Wohlfarth-Bottermann. *Cytobiologie* **18**, 76. (1978).
49. T. Ueda and Y. Kobatake. *Europ. J. Cell Biol.* **23**, 37. (1980).
50. T. Ueda, H. Satoh and Y. Kobatake. *Europ. J. Cell Biol.* **25**, 212. (1981).
51. T. Ueda, S. Koya and Y. Maruyama. *BioSystems* **54**, 65. (1999).
52. T. Hirose, T. Ueda and Y. Kobatake. *J. Gen. Microbiol.* **128**, 2647. (1982).
53. H. Satoh, T. Ueda and Y. Kobatake. *Cell Stru. Func* **9**, 37. (1984).
54. Y. Mori, T. Ueda and Y. Kobatake. *Protoplasma.* **139**, 141. (1987).
55. T. Ueda. *Sensors and Materials* **7**, 1. (1995).
56. G. Nicolis and I. Prigogine. *Self-organization in nonequilibrium systems*, (John Wiley & Sons Inc., New York, 1977).
57. T. Ueda, K. Matsumoto, T. Akitaya and Y. Kobatake. *Exp. Cell Res.* **162**,

486. (1986).
58. Y. Mori, K. Matsumoto, T. Ueda and Y. Kobatake. *Protoplasma* **135**, 31. (1986).
59. T. Ueda, Y. Mori and Y. Kobatake. *Exp. Cell Res.* **169**, 191. (1987).
60. T. Ueda, T. Nakagaki, and Y. Kobatake. *Protoplasma Sup.* **1**, 51. (1988).
61. T. Ueda, T. Nakagaki and T. Yamada. *J. Cell Biol.* **110**, 1097. (1990).
62. T. Ueda. *Phase Transitions* **45**, 93. (1993).
63. T. Ueda. *Forma* **13**, 405 .(1998).
64. T. Nakagaki, H. Yamada and A. Toth. *Nature* **407**, 470. (2000).
65. T. Nakagaki, H. Yamada and T. Ueda. *Biophys. Chem.* **84**, 195. (2000).
66. T. Nakagaki, R. Kobayashi, Y. Nishiura and T. Ueda. *Proc.R..Soc. Lond.* **B, 271**, 2305. (2004).
67. T. Nakagaki, S. Umemura, Y. Kakiuchi and T. Ueda. *Photochem. Photobiol.* **64**, 859. (1996).
68. Y. Kakiuchi and T. Ueda. *Protoplasma* **206**,131. (1999).
69. Y. Kakiuchi, T. Takahashi, A. Murakami and T. Ueda. *Photochem. Photobiol.* **73 (3)**, 324. (2001).
70. T. Ueda. *Trends in Photochem. Photobiol.* **1**, 25. (1990).
71. K. Suzuki and T. Ueda, *at Ann.Meeting of Biophys. Soc.* (2002).
72. K. Takahashi and T. Ueda. *at Ann.Meeting of Biophys Soc.* (2003).
73. K. Matsumoto and T. Ueda. In *Physics of Living State: Dynamics of Information and Fluctuation in Biological Systems*, ed. K. Sawada, T. Musha, T. Kawakubo, H. Shimizu (Ohmusha Ltd., 1994), p. 123.
74. S. Koya and T. Ueda. *ACH Models in Chemistry*, **135 (3)**, 297. (1998).
75. T. Ueda and Y. Kobatake. *J. Theor. Biol.* **97**, 87. (1982).
76. Y. Kakiuchi and T. Ueda. In *Circadian Organization and Oscillatory Coupling*, eds. K.Honma & S.Honma, (Hokkaido University Press, 1996), p. 79.
77. T. Ueda and Y. Kobatake. *Exp. Cell Res.* **147**, 466. (1983).
78. H. Satoh, T. Ueda and Y. Kobatake. *Exp. Cell Res.* **156**, 79. (1985).
79. T. Akitaya, S. Ohsaka, T. Ueda and Y. Kobatake. *J. Gen. Microbiol.* **131**, 195. (1985).
80. T. Akitaya, T.Ueda and Y. Kobatake. *Cell Struc. Func.* **11**, 1. (1986).
81. http://www.genome.gov/12511858
82. V.A. Vasilev, Yu. M. Romanovski, D. S. Chernaviskii, V. G. Yakhno. *Autowave Processes in Kinetic Systems*, (VEB Deutscher Verlag der Wissenschaften, 1987).

CHAPTER 10

COMMUNICATION AND STRUCTURE WITHIN NETWORKS

Kim Sneppen, Martin Rosvall, Ala Trusina

Models of Life center, Niels Bohr Institute,
Blegdamsvej 17, 2100 Copenhagen Ø, Denmark
NORDITA, Blegdamsvej 17, 2100 Copenhagen Ø, Denmark
E-mail: sneppen@nbi.dk

Life without information is not life. From the genetic blueprint in our DNA to the world-wide Internet, information and its dynamic counterpart communication shape our world. However, we live under the limited information horizon, in the sense that information is often imperfect and communication is always finite. Complex networks opens for a quantitative frame to discuss these issues, both through model building, through comparative studies of topologies and through studies of signaling in real networks. In this paper we explore a model for evolution of social networks, and discuss the general interplay between communication, specific signaling and overall topologies in various complex networks.

10.1. Introduction

The interactions in complex systems are typically associated to exchange of information and energy. Examples of complex systems where these interactions are fundamental and inherent properties, inseparable from the remaining system are: A living cell with its metabolic and signaling network, the ecosystem of co-evolving life formed on earth including our surrounding society. The emerged arena with complex networks have proven to contribute with useful methods to characterize systems that are composed of parts with individuality. The individuality is an important aspect, as it emphasizes the importance to consider the systems as complex networks where each part only interacts with a few other parts. In this way all parts of the complex network can be reached from other parts, but distant communication is obviously neither as easy nor as accurate as close direct communication [1]. That is, a network is de facto a description of the limited ability to send signals in the corresponding complex system [2]. Thus the study of networks is first of all a study of the relative importance or ease of communication ability at different distances. In this paper we therefore 1) discuss how network topology may be shaped dynamically under a constrained transfer of information[2], and 2) investigate how the topology of networks in general constrain the ability to send specific signals across the network [3,4].

10.2. An Economy for Exchange of Social Contacts

We here first consider a dynamic network that consists of agents that each tries to optimize their position, given their limited information. A natural quantity to optimize is the access to the overall system/network. That is, this social game is a game where individuals try to optimize their centrality in the network, and thereby gain information about other agents in the network, which they in turn can make use of. Let us further consider a scenario where links are not cost free, implemented by a dynamics constrained by a fixed total number of edges (and nodes).

Given the fixed number of edges, the hub structure is the core of any network where distances are truly minimized [5]. In fact, in such a simple centralized system where all agents (except the center) have only one link, any agent is only at max two steps away from any other agent. Thereby the information associated to locating a given other agent is in principle small. It may however be difficult to sustain such a centralized structure if each node has only limited information about the location of other vertices. In that case, when changing their neighbors by moving links from one node to another, they may make mistakes. This will destabilize the centralized hub, and may lead to a distributed network and a less optimal geometry.

To study the interplay between information exchange and dynamical rewiring of edges in a network, we introduce a simple agent based model [2] where different agents have different and adjustable memories, see Fig. 10.1. Every agent, named by a number $i = 1, 2, 3 \ldots n$, is a node in a connected network that consists of N nodes and E edges. Agent i has a memory

$$M_i = \begin{cases} D_i(l) \\ P_i(l) \end{cases}, \; l = 1, 2, \ldots, i-1, i+1, \ldots N,$$

with $N - 1$ *distances* D and *pointers* P to the other agents in the network. The distance $D_i(l)$ is agent i's estimated shortest path length to l. The pointer $P_i(l)$ is agent i's nearest neighbor on the estimated shortest path

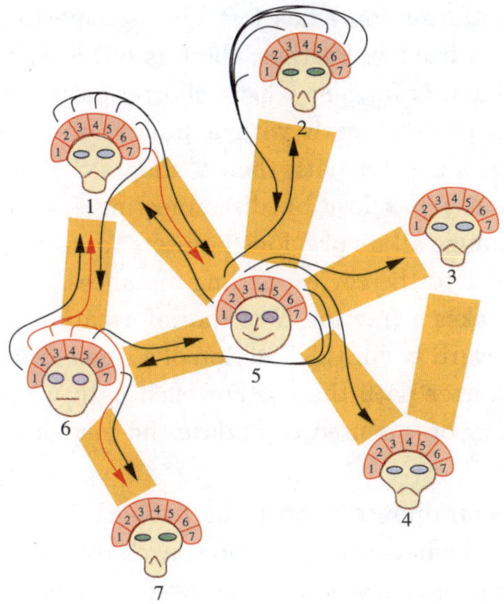

Fig. 10.1. Illustration of the agent based model, where information about other agents is used to decide where one should move links (the thick shaded boxes). In turn, the links are used to communicate with neighbors and thereby obtain a more correct picture of the placements of other agents. When communication rate is high compared to rate of moving links, a centralized system emerges. When communication rate is low, the system decentralizes and the network becomes chaotic and develops a narrow degree distribution. The black arrows indicate cases where agents have the right picture of other agents' position, the red arrows indicate false pictures.

to l. Thus M_i may be seen as a simplified version of the gateway protocol used by the autonomous systems to direct transmission of E-mails across the hardwired internet. Here, however, the memory will be used to rewire edges in the network.

Initially the network is a hub of the $N-1$ agents connected to a center agent by $N-1$ edges (as in Fig. 1, left) plus $E-N+1$ randomly placed edges on the periphery of the "hub". The rewiring consists of three steps:

(i) An agent i and one of its neighbors j is chosen randomly.
(ii) An agent $l \neq i$, j is chosen randomly and if $D_i(l) > D_j(l)$ then the edge between i and j is rewired to an edge between i and $k = P_j(l)$. If l did not satisfy the above criteria a new l is randomly chosen. If no such l exists the rewiring is aborted.

(iii) The information i has lost by disconnecting j is replaced by information from k. Further, there is full exchange of information between i and k: If agent k list a shorter path to some other agents, then i adopts this path with a pointer to k. Similarly for k, if agent i list a shorter path then k adopts this path through i. The information j has lost by disconnecting i is replaced by letting agent j change all its previous pointers toward i to pointers toward k and add 1 to the corresponding distances.
(iv) One takes a fraction S of the information j has stored with a pointer toward k. For this fraction of j's memory one check whether k list a shorter path than j. For each path where this is the case, the memory of k is used to update the memory of j.

The model defines an update of both the network (ii) and the information (iii) that agents in the network have about each others' locations. The step (ii) represents local optimization where agent i rewire from j to k with a probability given by the fraction of the network which is estimated to becomes closer. Notice that only a small part of the system is informed about a changed geometry, and that decisions on moves may be based on outdated information. The parameter S in step (iv) represent the level of communication in the network, as it quantifies the information exchange between agent j and agent k. Notice also that the update in (iv) takes place no matter which node had the right data. When $S = 0$ the network develops a fairly narrow degree distribution, whereas $S = 1$ allows self organization to single a hub like structure reminiscent of a monarchy. In between there is a critical value of $S = S_{crit} \sim 0.1$ (for $\langle C \rangle = 3$) where one obtains a scale free degree distribution, see [2]. S_{crit} depends on the overall edge density in the system, and increases as the average degree $\langle C \rangle$ increases. To maintain the central hub one should minimize the number of links in the system, with correct information along the few links that all lead to the central hub. This is a "divide and rule" strategy.

In any case, at conditions when one hub dominates the topology, the center of this hub becomes frozen at the initial hub. Thus, a necessary condition for a scale-free degree distribution is instability of large hubs and the possibility for vertices to change status dynamically. On the other hand, when the instability becomes to large, no large hubs develop and the degree distribution becomes Poisson-like. In sociological terms, a scale-free degree distribution emerge when the gain of monopoly balance its inherent imperfection.

10.3. Limited Information Horizons in Complex Networks

The presented "economic" model of link exchange between "egoistic" agents with limited access to information, teaches us about communication ability and positioning in networks. A philosophy that to some extent can be used to investigate networks which are formed of entirely different nature than the social games between humans. In fact, in more general terms we want to demonstrate that one can characterize different network topologies in terms of their ability to facilitate *specific* communication [3,4].

That is, we characterize how the topology of networks constrain the ability to send specific signals from one specific- to another specific node. We stress the fundamental difference between specific signaling and non-specific broadcasting: Where the first only focuses on locating one specific node without disturbing the remaining network, the broadcasting amplifies by transferring the signals to all exit links of every node along all branching paths. Specific signaling is thus constructive, whereas broadcasting rather is of relevance for disease spreading, spam, or computer virus propagation. Networks are the result of specific interactions, whereas the non specific communication rather is associated to a cost that in principle should be minimized.

This view of networks can be formalized in terms of information measures that quantify how easy it would be for a node to send a signal to other specific nodes in the rest of the network [3,4]. To do this one count the number of bits of information required to *transmit* a message to a specific remote part of the network, or reversely, to *predict* from where a message is received (see Fig. 10.2).

In practice, imagine that you at node i want to send a message to node b in a given network (left panel in Fig. 10.2). Assume that the message follows the shortest path. That is, as we are only interested in specific signals we limit ourselves to consider only this direct communication. If the signal deviate from the shortest path, it is assumed to be lost. If there are several degenerate shortest paths, the message can be sent along any of them. For each shortest path we calculate the probability to follow this path, see Fig. 10.2. Assume that without possessing information one would chose any new link at each node along the path with equal probability. Then:

$$P\{p(i,b)\} = \frac{1}{k_i} \prod_{j \in p(i,b)} \frac{1}{k_j - 1}, \tag{1}$$

where j count all nodes on the path from a node i to the last node before the target node b is reached. The factor $k_j - 1$ instead of k_j takes into account

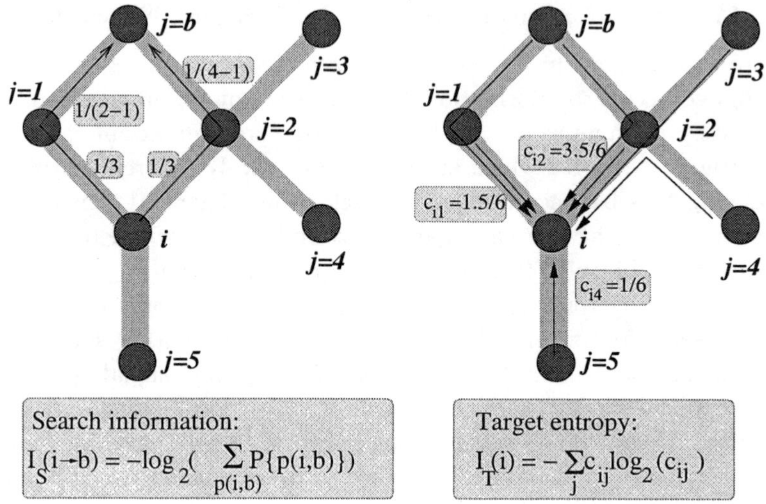

Fig. 10.2. Information measures on network topology: Left panel: *Search information* $S(i \to b)$ measures your ability to locate node b from node i. $S(i \to b)$ is the number of yes/no questions needed to locate any of the shortest paths between node i and node b. For each such path $P\{p(i,b)\} = \frac{1}{k_i} \prod_j \frac{1}{k_j - 1}$, with j counting nodes on the path $p(i,b)$ until the last node before b. Right panel: *Target entropy* T_i measures predictability of traffic to node i. c_{ij} is the fraction of the messages targeted to i that passed through neighbor node j. Notice that a signal from b in the figure can go 2 ways, each counting with weight 0.5.

the information we gain by following the path, and therefore reduce the number of exit links by one. In Fig 10.2 we show the subsequent factors in going along any of the two shortest path from node i to node b. The minimal information needed to identify one of all the degenerate paths between i and b defines the "search information"

$$I_S(i \to b) = -log_2 \left(\sum_{p(i,b)} P\{p(i,b)\} \right), \tag{2}$$

where the sum runs over all degenerate paths that connect i with b. A large $I_S(i \to b)$ means that one needs many yes/no questions to locate b. The existence of many degenerate paths will be reflected in a small I_S and consequently in easy goal finding.

The value of $I_S(i \to b)$ teaches us how easy it is to transmit a specific message from node i to node b. To characterize a node one may ask how

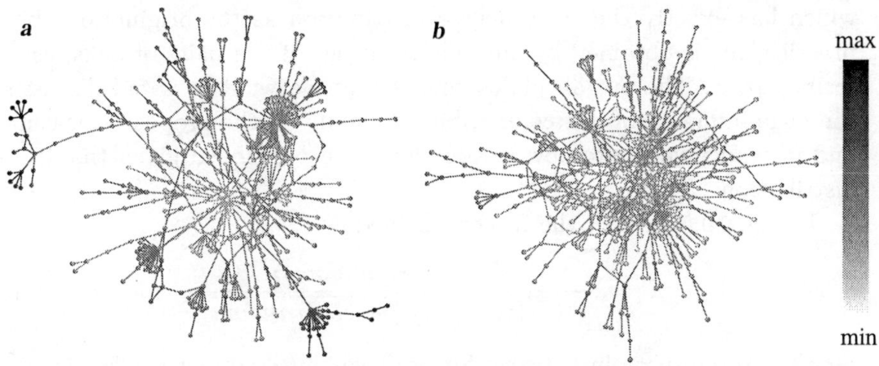

Fig. 10.3. **a)** Analysis of the protein-protein interaction network in yeast defined by the connected component of the most reliable data from the two-hybrid data of Ito et al. 2001. The value of the shown access information $\mathcal{A}_i$ is increasing from being light colored in center to darker in periphery. The dark colors mark nodes that have least access to the rest of the network, i.e. proteins that are best hidden from the rest. **b)** show a randomized version of the same network. One see that hubs are more interconnected and that typical $\mathcal{A}$ values are smaller (less dark).

easy is it in average to send a specific message from one node in the net:

$$\mathcal{A}(i) = \sum_b I_S(i \to b) \tag{3}$$

$\mathcal{A}$ is called the access information. In Fig. 10.3 show $\mathcal{A}(i)$ for proteins belonging to the largest connected component of the yeast protein-protein interaction network obtained by two hybrid method. The shown network nicely demonstrate that often highly connected nodes are on the periphery of the network, and thus do not provide particular good access to the rest of the system [6]. This is not what one see in a randomized version of the network (right panel). In fact we quantify the overall ability for specific communication

$$\mathcal{I}_S = \sum_i \mathcal{A}(i) = \sum_{i,b} S(i \to b) \tag{4}$$

and compare with the value $\mathcal{I}_S(random)$ obtained for a randomized network.

We stress that it is very important to compare with a meaningful randomized network, that is a network which is consist of one component and which has exactly the same degree distribution as the original one [6]. In practice this is obtained by multiple rewirings of the pairs of links, as described by [6]. The overall philosophy is that we want to avoid discussing the huge influence of degree distribution on any of the proposed measures, and instead investigate how various hubs are positioned given the degree distribution [7].

In Fig. 10.5 we plot the Z score defined as

$$Z = \frac{\mathcal{I}_S - \langle \mathcal{I}_S(random)\rangle}{\sqrt{\langle \mathcal{I}_S(random)^2\rangle - \langle \mathcal{I}_S(random)\rangle^2}} \tag{5}$$

for the protein-protein network for both yeast (*Sacromyces Cerevisia*) [8,9] and fly (*Drosophilia*) [10] as well as for the hardwired Internet [11] and a human network of governance (CEO) defined by company executives in USA where two CEO's was connected by a link if they are members of the same board [12]. One sees that $\mathcal{I}_S > \mathcal{I}_S(random)$ for all real world networks. Thus networks have topologies which tends to hide nodes from each other on average.

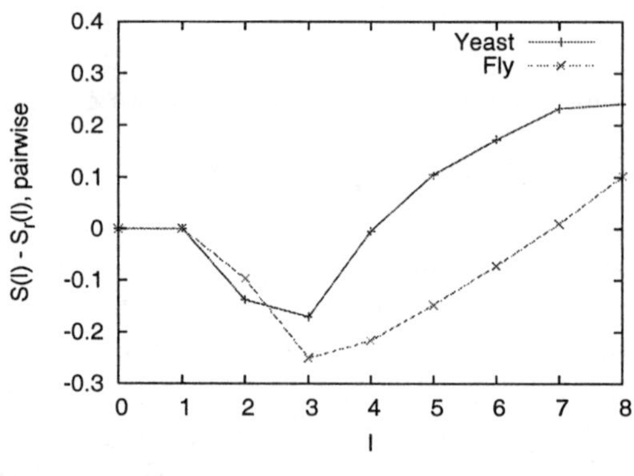

Fig. 10.4. Average information needed to send a specific signal to a node at distance l, compared to the information needed if the network was random. Both cases refer to the protein-protein interaction networks, measured by the two hybrid method. In both cases the random network is constructed such that for each protein, its number of bait and its number of prey partners are conserved during the randomization.

The tendency to hide, respective communicate, can be quantified further by considering the average information $\langle S(l) \rangle$ needed to send a specific signal a distance l inside the network (the average is over all nodes and all neighbors at distance l to these nodes in the given network). This is done in Fig. 10.4. We see that $\langle S(l) \rangle - \langle S_{random}(l) \rangle$ have a minimum below zero for some rather short distance $l \sim 3$, whereas it becomes positive for large l. Thus the molecular signaling networks have relatively optimal topology for local specific communication, but on larger distance the proteins tend to hide.

In Fig. 10.5 we also show another quantity, the ability to predict from which of your neighbors the next message to you will arrive from. This quantity measures predictability, or alternatively the order/disorder of the traffic around a given node i. The predictability based on the orders that are targeted *to* a given node i is

$$I_T(i) \;=\; -\sum_{j=1}^{k_i} c_{ij} log_2(c_{ij}), \tag{6}$$

where $j = 1, 2..., k_i$ denotes the links from node i to its immediate neighbors j and c_{ij} is the fraction of the messages targeted to i that passed through node j. As before our measure is implicitly assuming that all pairs of nodes communicate equally with one another.

Notice that I_T is an entropy measure, and as such is a measure of order in the network. In analogy with the global search information $\mathcal{I}_S$ one may also define overall predictability of a network

$$\mathcal{I}_T = \sum_i I_T(i) \tag{7}$$

and compare with its random counterparts. In general as the organization of a network gets more disorganized $\mathcal{I}_T$ increases and the number of hubs with disordered traffic increases. Also, as one consider networks with increasing value of $\mathcal{I}_T$, nodes of low degree tend to be positioned between the hubs. In this way most links from most nodes will be equally useful in reaching other parts of the network, representing a robust and also highly disordered network topology. In contrast a low $\mathcal{I}_T$ will be highly polarized in the sense hubs are positioned hierarchically and such that only the few links between them are effectively used.

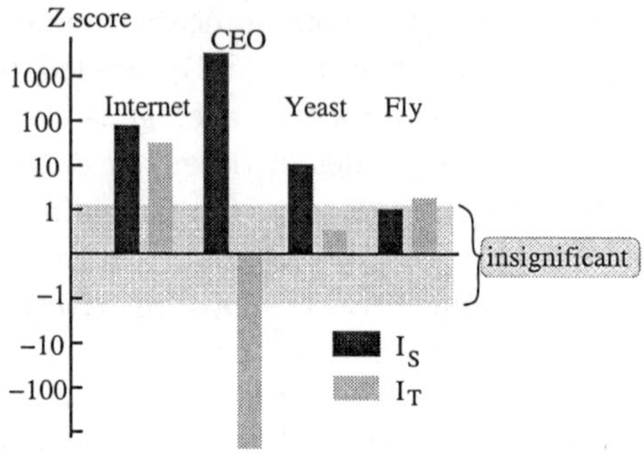

Fig. 10.5. Measure of communication ability of various networks [3]. A high Z-score implies relatively high entropy. In all cases we show $Z = (\mathcal{I} - \mathcal{I}_r)/\sigma_r$ for $\mathcal{I} = \mathcal{I}_S$ and $\mathcal{I}_T$, by comparing with $\mathcal{I}_r$ for randomized networks with preserved degree distribution. σ_r is the standard deviation of the corresponding $\mathcal{I}_r$, sampled over 100 realizations. Results within the shaded area of two standard deviations are insignificant. All networks have a relatively high search information $\mathcal{I}_S$. The network of governance CEO show a distinct communication structure characterized by local predictability, low $\mathcal{I}_T$, and global inefficiency, high $\mathcal{I}_S$.

10.4. Conclusion

Many networks are coupled to specific communication and their topology should reflect this. The optimal topology for information transfer relies on a system-specific balance between effective communication (*search*) and not having the individual parts being unnecessarily disturbed (*hide*). For molecular networks of both yeast and fly we observed that communication is good at short distances, whereas it is relatively bad at large distances. We found similar search features in the hardwired Internet and information city networks [13]. This overall feature was found when comparing the real network with its randomized counterpart given the degree sequence. However, we have also found that scale-free networks in turn have a larger $\mathcal{I}_S$ than the totally randomized network where only the average degree is conserved. Thus also the broad degree distribution found in many real world networks makes long-distance communication more difficult: $\mathcal{I}_S(real, scale-free) > \mathcal{I}_S(random, scale-free) > \mathcal{I}_S(random, Erdos-Reynei)$.

The generic feature, found also for other investigated networks [13,14], is that for real networks $\mathcal{I}_S$ is larger than its random expectation. This presumably reflects an overall tendency to avoid long distance non-specific communication. Instead we found that networks are constructed to favor communication at short distances, where local inter-connected neighborhoods develop a topology that protects against long-distance disturbance, but favours short-distance communication. Whereas $\mathcal{I}_S$ cannot be used to classify networks qualitatively, we suggested the target entropy $\mathcal{I}_T$ that for example differentiate between the relatively predictable networks of human governance and the more robust structure of for example the Internet.

Acknowledgments

We acknowledge support of Swedish Research Council through Grants No. 621 2003 6290 and 629 2002 6258 and from the Danish "Grundforskningsfond".

References

1. N. E. Friedkin, The University of North California Press (1983).
2. M. Rosvall and K. Sneppen, Phys. Rev. Lett. **91**, 178701 (2003).
3. K. Sneppen, A. Trusina and M. Rosvall, cond-mat/0407055 (2004). To appear in Europhysics letters.
4. M. Rosvall, A. Trusina and K. Sneppen, Phys. Rev. Lett. **94**, 028701. cond-mat/0407054 (2004).
5. A. Valverde, R. Ferreri Cancho, R.V. Sole, cond-mat/0204344 (2002).
6. S. Maslov and K. Sneppen, Science **296**, 910 (2002).
7. A. Trusina, S. Maslov, P. Minnhagen & K. Sneppen, Phys. Rev. Lett. **92**, 178702 (2004).
8. P. Uetz *et al.* Nature **403**, 623 (2000).
9. T. Ito *et al.* Proc. Natl. Acad. Sci. U.S.A. **98** 4569 (2001).
10. Giot, L. *et al.* Science **302** 1727 (2003).
11. Website maintained by the NLANR Measurement and Network Analysis Group at http://moat.nlanr.net/
12. G. F. Davis and H. R. Greve, American Journal of Sociology **103**, 1 (1997).
13. A. Trusina, M. Rosvall, K. Sneppen. "*Information Horizons in Networks*". cond-mat/0412064.
14. M. Rosvall, P. Minnhagen, K. Sneppen. "*Navigating Networks with Limited Information*". cond-mat/0412051.